RAJA BEN AHMED LAJILI

The Hirudinées of the Lebna Dam: Biodiversity and Perspectives

RAJA BEN AHMED LAJILI

The Hirudinées of the Lebna Dam: Biodiversity and Perspectives

Discovering little-known fauna and new species in Tunisia

ScienciaScripts

Imprint
Any brand names and product names mentioned in this book are subject to trademark, brand or patent protection and are trademarks or registered trademarks of their respective holders. The use of brand names, product names, common names, trade names, product descriptions etc. even without a particular marking in this work is in no way to be construed to mean that such names may be regarded as unrestricted in respect of trademark and brand protection legislation and could thus be used by anyone.

Cover image: www.ingimage.com

This book is a translation from the original published under ISBN 978-620-6-69867-8.

Publisher:
Sciencia Scripts
is a trademark of
Dodo Books Indian Ocean Ltd. and OmniScriptum S.R.L publishing group

120 High Road, East Finchley, London, N2 9ED, United Kingdom
Str. Armeneasca 28/1, office 1, Chisinau MD-2012, Republic of Moldova, Europe
Printed at: see last page
ISBN: 978-620-8-31391-3

Contents

INTRODUCTION

Annelids represent an important step in the revolution of the animal world. In fact, they were the first coelomate triploblastic metazoans. They are divided into three classes: Polychetes, Oligochetes and Achetes. At present, some authors accept that the Oligochetes and Achetes are two subclasses of the Clitellates.

The Achetes, also known as Hirudinees, are the most evolved of the Annelids and have affinities with the Oligochetes. They are known by the common name of "leeches", which refers to the hematophagous species that are the best known and used in medicine.

Knowledge of leeches dates back to ancient times. They are mentioned in the Bible, in the thirtieth chapter of Proverbs and in the Koran; it seems that the word "Aluka" refers to the leech. Several Arab authors have also named this annelid "Aleka", "Aletha" and "Aleg" **(Grasse, 1959).** Traces of these annelids, Achetes, have been found dating back to the Jurassic era, such as *Epitrachys rugosus* and *Palaeohirudo eichstaettensis* **(Siddall *et al.*, 2001).**

Currently, 650 species of Hirudinae have been described worldwide; in fact, this figure is underestimated and there are probably still a number to be discovered **(Borda and Siddall, 2004 b).**

Several monographs have been devoted to the systematic study of Hirudineae worldwide, including those on the Hirudineae of Italy **(Minelli, 1979),** North America **(Blanchard, 1917) and** Australia **(Govedich, 2001),** not forgetting the most classic monograph **(Grasse, 1959)** and the numerous works by **Soos (1966, 1967, 1968 and 1969)**. With regard to the systematic position of the species collected, we will refer here to the most complete monograph on this group, by **Sawyer (1986).**

More recently, many researchers around the world have been interested in the study of leeches. These include work on systematics and phylogenetics **(Govedich and Davies, 1998; Govedich *et al.*, 1998; Govedich *et al.*, 1999; Govedich, 2001; Govedich *et al.*, 2002; Siddall, 2001 a and b; Siddall *et al,* 2001; Siddall, 2002; Siddall and Borda, 2003 and 2004; Westergren and Siddall, 2004; Borda and Siddall, 2004 a and b; Siddall and Budinoff, 2005; Siddall *et al.*, 2005; Utevsky and Trontelj, 2005; Figueroa and Regagnon, 2005; Figueroa *et al*, 2005 and Figueroa , 2006)**, reproduction **(Gullo, 1994, 2003, 2004, 2006 and Govedich *et al.*, 2003)**, developmental biology **(Shankland, 1990, 1994 and 1995; Denise and Shankland, 1992; Shankland and Seaver, 2000; Seaver and Shankland, 2000 and 2001)**, physiology **(Esch and Kristan, 2002; Kristan *et al*, 2005 and Burgin *et al.*, 2006)** and ecology **(Kutschera and Wirtz, 2001; Govedich *et al.*, 2004; Kutschera and Roth, 2005 and 2006 a and b).**

In Tunisia the fauna of the Hirudinees has hitherto been the subject of only a few occasional reports; on this occasion we stress the absence of determination keys for the leeches of North Africa. **Megnin (1891)** had observed in Paris, in the oral cavity of horses brought back from Tunisia by French soldiers, a leech which he had determined to be *Haemopis sanguisuga* (Linne, 1785) and which in fact turned out to be *Limnatis nilotica* (Savigny, 1820) **(Moore, 1939). Blanchard (1981)** mentioned the following species from the Tunisian fauna: *Batracobdella algira* (Moquin-Tandon, 1846), observed on the epidermis of various species of anuran (*Discoglossus pictus*, *Rana rudibunda*) and urodele (*Hydromantes genei*) amphibians, *Placobdella costata* (Fr. Muller, 1846) (under the incorrect name of *Haementeria costata*), *Dina lineata* (O.F. Muller, 1774) under the name of *Dina quadriseriata* (Grube, 1850), *Limnatis nilotica* (Savigny, 1820) and *Hirudo troctina* (Johnson, 1816). **Seurat (1922)** cited *Hirudo troctina* from the whole of North Africa, and more particularly from Algeria and Tunisia, *Batracobdella algira* (Moquin-Tandon, 1846) mentioned erroneously by **Viguier (1879)**, thinking he was dealing with a new species under the name *Batracobdella latasti* (Viguier, 1879). **Nesemann and Neubert (1994)** reported the species *Trocheta africana* for the first time in Tunisia. Until now, it has only been known from our country. According to

According to Nesemann and Neubert (1994), this species could correspond to *Dina quadriseriata* (Grube, 1850) mentioned by **Blanchard (1981)** for Tunisia. Finally, **Hetchel and Sawyer (2002)** confirmed the presence of *Limnatis nilotica* and *Hirudo troctina* in Tunisia, and reported new Tunisian localities of the latter based on the identification of as yet undetermined specimens held in the collections of the Museum national d'histoire naturelle in Paris.

The aim of this study, which focuses on the biodiversity of Annelids Achetes in the Lebna dam, as well as their ecological preferences and biometric characteristics, is to :

- To shed light on a small part of the annelid fauna of Achete tunisienne and thus contribute to the work of **Sawyer (1986)** concerning the region of North Africa.
- To reveal the ecological preferences of the species surveyed and their seasonal distribution.
- Suggest the biological cycles of the leech species collected.

This work comprises two parts, the first of which is divided into two chapters: - The first deals with the general morphological, anatomical and systematic characteristics of the Hirudinae. These general descriptions were used in the study of the species collected and their identification.

- The second chapter is devoted to a presentation of the study environment, the method for collecting Hirudineae and the experimental protocols used.

The second section, on results, is divided into two chapters:

- Chapter I is devoted to a systematic study of the different species found in the Lebna dam reservoir.

- In Chapter II we present the variations that the abiotic factors of the study environment can undergo and their influences on the dynamics of the populations studied. We have also tried to estimate the oviposition period of some of the species collected.

CHAPTER I

GENERAL INFORMATION ON HIRUDINES

I. DEFINITION

Annelids are triploblastic metazoans with bilateral symmetry. The mesoderm is hollowed out into closed vesicles: they are therefore coelomates; the body is formed by a succession of identical elements, the metameres. The blastopore coincides with the mouth, so they are protostomans.

The Annelid phylum is subdivided into 3 classes:

- The Polychetes class, characterised by numerous bristles.
- The Oligochetes class is characterised by sparse bristles.
- The Achetes class is characterised by the absence of bristles.

The Achetes, still called Hirudinees, are generally considered to be the most highly evolved Annelids. They have affinities with the Oligochetes. They are aquatic, freshwater or marine organisms, frequently ectoparasites of vertebrates, commonly known as leeches. Some exotic species are terrestrial and haunt damp areas.

These worms generally have a front suction cup pierced through the mouth for suction and fixation, and a posterior suction cup which is a simple ventral fixation device in relation to the anus.

These animals are hermaphrodites. The eggs laid are rich in vitellus. They undergo direct development, so there is no trochophore larva like the other two classes.

II. GENERAL ORGANISATION

1. Tegument

The body of leeches is entirely covered by a unistratified epithelium comprising common epithelial cells, isolated or grouped sensory cells and mucous-type glandular cells **(Brumpt, 1949)**. There are also unicellular clitellar glands with multi-lobed nuclei. The latter are specialised in cocoon secretion and are of two types: some secrete a chitinous substance which hardens and forms the cocoon envelope, while the others produce an albuminous liquid which is used to feed the embryo.

2. Musculature

Hirudinees have a very powerful musculature; the body is in fact highly contractile: an individual such as *Haemopis sanguisuga*, measuring 12cm, can contract by up to 2.5cm under the action of a fixative. The musculature of leeches is made up of smooth muscle fibres of variable length, and an outer layer of longitudinal muscles that run the length of the body and converge towards the extremities to end in the suction cups.

Between the circular and longitudinal muscles is a layer of oblique muscles. There are also dorso-ventral muscles, which are generally short **(Grasse *et al.*, 1970)**.

3. Metalwork and body regions

The leech body has a defined and constant number of metameres or somites, equal to 34. This constant number is explained by the absence of postembryonic pygidial growth. Externally, there are 101 visible rings for 34 metameres.

In *Hirudo medicinalis*, metameres I, II, III, IV, V and VI correspond to 1, 1, 2, 3, 3 and 3 rings respectively. In the middle region of the body, each segment is subdivided into five rings. The preoral lobe and ventral portion of the first five metameres contribute to the formation of the buccal sucker; the posterior sucker results from the transformation of the last seven metameres (XXVII to XXXIV) **(Beaumont et Cassier, 1981).**

Four elements can be used to limit a metamer:

- The metameric position of the nephridiopores: a slight pressure on the animal causes a drop of liquid to be ejected from the nephridiopores.
- The position of the sensory papillae: These are visible on the dorsal surface of the leeches and are positioned on the central ring of each metamer.
- The metameric repetition of rings that are larger or smaller than their neighbours.
- The metameric repetition of coloured spots or stripes.

The body of a leech is divided into 5 regions **(Minelli, 1977)** (Fig. 1) :

3.1. Cephalic region

It is clearly delimited, with a protomerite (which corresponds to the prostomium of other Annelids) and four somites. The eyes are situated dorsally and the anterior sucker, into which the mouth opens, is situated ventrally.

3.2. Pre-Clitellian region

It is short and made up of four somites containing the proboscis, salivary glands and cerebral ganglia.

3.3. Clitellian region

It is also known as the sexual belt and is made up of a number of segments located close to the genital openings. These segments become intumescent during the reproductive period. The male orifice opens on the medio- ventral line of somite X, the female orifice on the medio- ventral line of somite XI. In *Callobdella laphii* the somite corresponding to the male orifice is coloured red. This region contains the clitellate glands which are essential during the oviposition period.

3.4. Average region

It generally shows normally developed somites in which the number of rings characteristic of the species can be counted. This region starts from somite XXVI; it contains the main viscera: circulatory system, nephridia, coelomic cavities and most of the digestive system.

3.5. Rear region

It is highly muscular, the result of the fusion of six somites, and is completely formed by the posterior sucker.

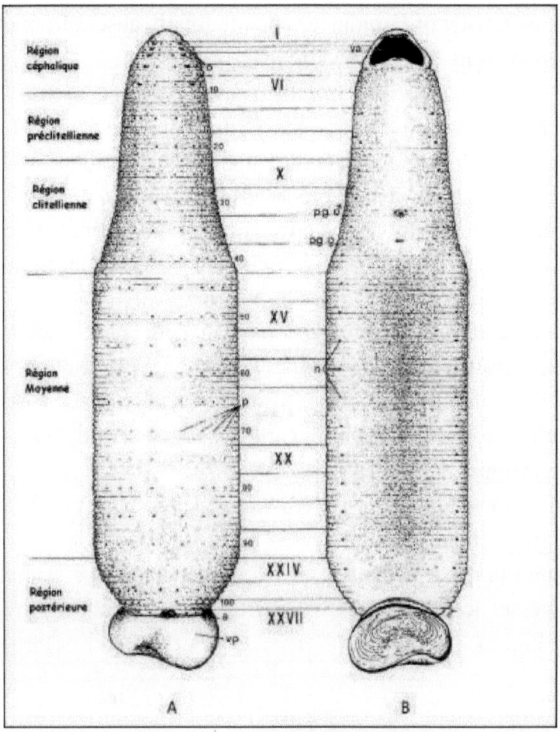

Fig. 1: Diagrams of the external morphology *of Haemopis sanguisuga.*
A **: dorsal view ; B : ventral view, va : anterior sucker, pg ♂: male pore, pg ♀: female pore, n : nephridiopores, vp : posterior sucker, a : anus, I to XXVII: somites, 1 to 101: rings (Minelli, 1979).**

4. Suction cups

These are characteristic of the morphology and physiology of Hirudinae. There are at most two suckers: one at the front, known as the buccal sucker, and the other at the rear, known as the terminal sucker **(Grasse, 1959).**

4.1. The front suction cup

Its shape enables it to exert suction on the surface to which it is applied. Its musculature is rarely powerful and is generally reduced to a single circular fibre and a few longitudinal muscles that taper before being inserted into the cuticle.

4.2. The rear suction cup

It is always discoid in shape, with its ventral side facing the ventral side of the body.

5. Connective tissue

A characteristic feature of Hirudinae is the presence of an overabundant connective tissue that reduces the general cavity to a coelomic lacunar network. This connective tissue, derived from primordial mesenchyme, comprises connective fibres, fat cells and pigmented cells.

Fat cells are characteristic of Hirudinees and are totally absent from other annelids.

6. Incomplete system

Typically, in *Glossiphonia complanata* the lacunar system consists essentially of :

- **a median lacuna** which crosses the body from the head to the anus, and divides into: a straight dorsal lacuna containing the dorsal blood vessel, and a larger ventral lacuna containing the ventral blood vessel, the nerve cord and the gonads.
- **lateral lacunae** running along the edges of the body, from the head to the anus. They are joined to the median lacuna by a circular channel that follows the edge of the 2 front and rear suckers.
- **intermediate lacunae** (between the median and lateral lacunae), which are considered to be expansions of the median lacuna between the 6eme somite and the anus. They form a continuous network of ducts occupying the space between the dorsal and ventral muscles, the nephridia and the connective tissue.

7. Vascular system

It varies according to whether they are Rhynchobdelles or Arhynchobdelles :

7.1. Rhynchobdelliformes

This blood system in a *Glossiphonia*, for example, consists essentially of a dorsal and a ventral vessel joined at the two extremities of the body by a number of vascular loops (Fig. 2).

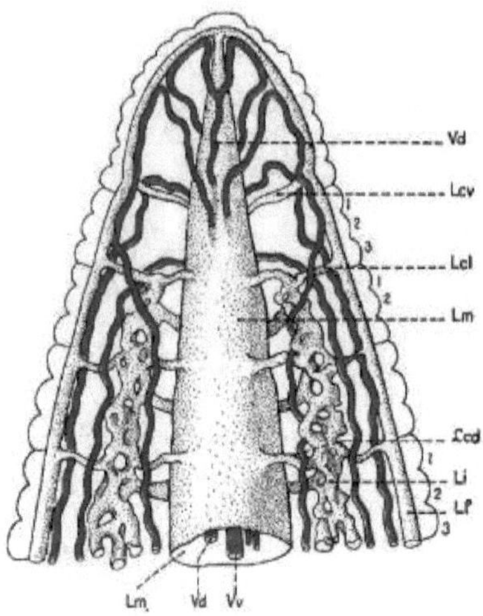

Fig. 2: Relationship between blood vessels (in red) and coelomic lacunae (in black) in *Glossiphonia complanata* in the anterior region of the body. The dorsal and ventral lacunae merge to form the median lacuna, which contains these vessels, both dorsal and ventral. Lcd, Lcl, Lcv: dorsal, lateral, ventral communication lacuna; Li: intercalary lacuna, Ll: lateral lacuna, Lm: median lacuna; Vd, Vv: dorsal and ventral blood vessels, 1, 2, 3: successive rings of the same segment (Grasse, 1959).

The blood of Rhynchobdellae is colourless or very faintly coloured; it is slightly pink in *Glossiphonia complanata.* In Glossiphonia complanata there is no hemoglobin of its own.

7.2. Arhynchobdelliformes

In *Erpobdella*, taken as an example of the Arhynchobdellae, the definitive vascular system comprises :

- A ventral vessel containing the nerve chain, bulges at the level of the ganglia.
- Two lateral vessels running from one end of the body to the other.
- Two fine capillary networks.
- Quite large vessels.
- Transverse vessels connect the ventral vessel with the lateral vessels.

The circulatory system of *Hirudo* has the same structural plan as that of *Erpobdella*, but with the addition of a dorsal vessel. The interstitial and subepidermal capillary networks are extremely dense. The blood of Arhynchobdellae is coloured red by haemoglobin dissolved in plasma.

8. Respiratory system

The Hirudineae, with a few exceptions, do not have a separate respiratory system; their gas exchange takes place through the epidermis. In Glossiphoniidae, respiration is essentially cutaneous. It is facilitated by the animal's rhythmic undulatory movements. For this function to take place, the epidermis must be moist. The poor development of cutaneous glands in the Clepsinidae, taken as an example of the *Glossiphoniidae*, explains their low resistance to air.

In the Erpobdellidae and Hirudinidae, a dense network of blood capillaries even extends between groups of epidermal cells. Various species of Hirudinae are able to use oxygen from the air and oxygen dissolved in water, as in the case *of Haemopis sanguisuga* and *Trocheta subviridis*.

9. Excretory system

Apart from the accumulating kidney, which in the most highly organised Hirudinae forms the botryoid tissue, all leeches have a segmental excretory system composed of a number of pairs of nephridia ranging from 10 to 17 pairs.

9.1. Nephridia

Initially, a nephridia (Fig. 3) consists of a blind capsule enclosing a phagocytic organ, the base of which is contiguous with the end of the nephridial tube proper. This capsule is surmounted by a funnel which places it in communication with the coelom. The nephridial tube is a very long circumferential excretory canal which turns back on itself several times and is pierced by a row of cells without cilia. It leads to a terminal bladder lined with ciliated epithelium and opening externally with a nephridiopore **(Grasse, 1959).**

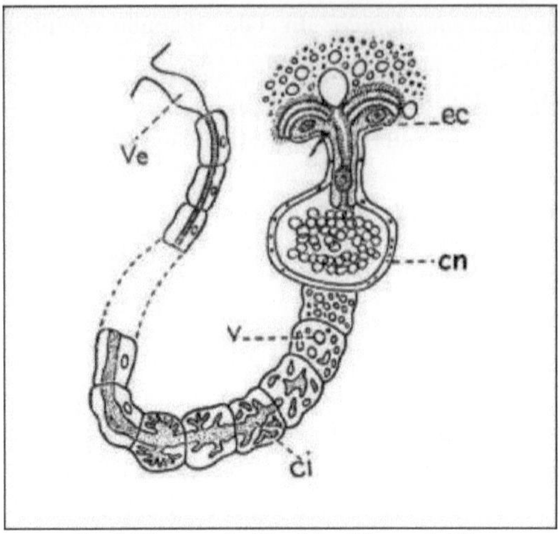

Fig. 3: Diagram of a Glossiphoniidae nephridia.
Ci: intracellular canal, ec: ciliated funnel cn: nephridial capsule, v: vacuoles with excretory granules, Ve: terminal bladder (Grasse, 1959).

10. Nervous system

The nervous system of leeches is based on the general type of that of Annelids, but is more developed than that of Oligochetes. The ganglia have a more sophisticated structure, probably due to the complication of the musculature.

There is a central nervous system, a peripheral nervous system and a sympathetic nervous system **(Grasse, 1959).**

10.1. Central nervous system

The central nervous system is made up of :

- **A peripharyngeal mass**: made up of cerebrospinal ganglia situated above the pharynx and ventral ganglia known as subpharyngeal ganglia. A long transverse commissure joins the cerebroid ganglia.
- **A ventral chain:** made up of isolated ganglia located behind the peripharyngeal mass. The neurons are contained exclusively in the follicles, inside the ganglia. The ventral chain comprises 27 or 28 neuromeres, ranging from neuromere VII to neuromere XXXIV.

Each ganglion has 6 neuronal follicles. From each side of the ganglion run 2 nerves: one anterior and one smaller posterior.

- **A posterior mass:** made up of the last 7 neuromeres (Fig. 4), the last of which, described as anal, is rudimentary. The nerves are arranged as in the ventral chain.

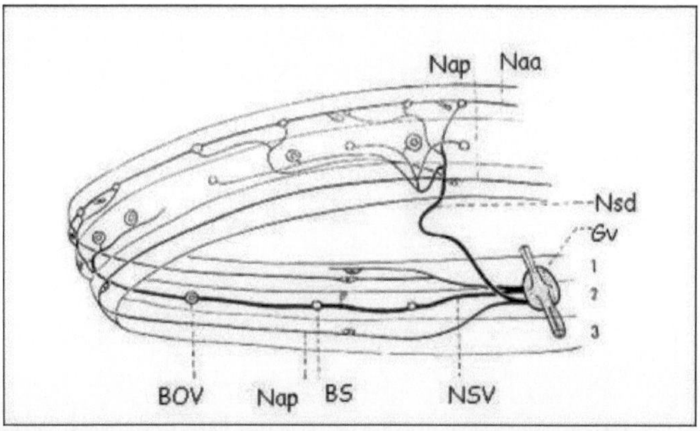

Fig. 4: Diagram of a neuromere from *Theromyzon tessulatum* 1, 2 and 3: somite rings, Bs: sensory bud, Bov: sensory bud with retinal cells, Gv: ventral ganglia, Naa, Nap: anterior and posterior annular nerves, Nsd: dorsal segmental or lateral nerve, Nsv: ventral segmental nerve (Grasse, 1959).

10.2. Peripheral nervous system

It is essentially made up of the nerves that run from the peripharyngeal mass, the ventral chain and the posterior nerve mass. These nerves branch and send their endings to the viscera and teguments.

10.3. Sympathetic or stomatogastric nervous system

From each side of the peripharyngeal collar runs a bundle of fibres, turned inwards, giving 3 nerve trunks: 1 dorsal, 1 lateral and 1 ventral. The latter has a small ganglion swelling. The fibres from these 3 sympathetic nerves branch out into a complex plexus, giving the image of a network; anteriorly, the sympathetic branch, which bifurcates into the dorsal and lateral nerves, sends bundles of fibres into the wall of the anterior pharynx.

11. Digestive system

Typically, the digestive tract of Hirudineae can be divided into three distinct regions: anterior, middle and terminal **(Doin, 1952)** (Fig. 5).

11.1. Previous region

It is covered along its entire length by a cuticle. Its morphology is used as a basis for the systematic division of leeches:

Rhynchobdelles: have a proboscis.

Arynchobdellae: lack a proboscis and have either two or three serrated jaws (Gnathobdellae) or three bulges, protruding longitudinally and muscularly (Pharyngobdellae).

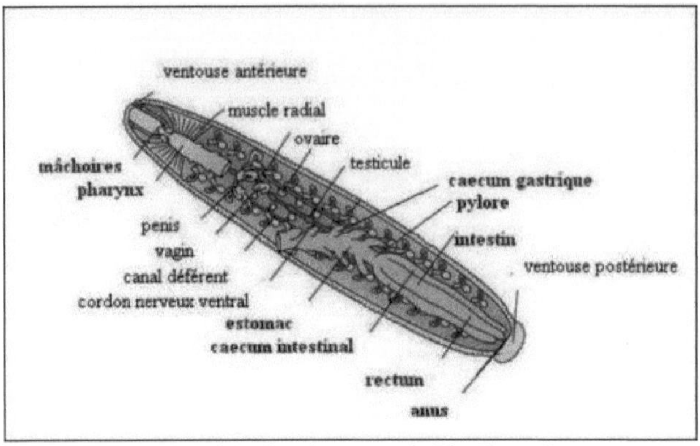

Fig. 5 : General organisation of the digestive system in Hirudinees (Anonymous1 : website).

This anterior region opens into a buccal orifice. This is a longitudinal slit cut into the prostomium in the ventral position. This mouth is followed by a vast cavity containing the proboscis.

The proboscis of Rhynchobdellae is itself a pharynx. It has a cylindrical shape (Fig. 6) and is extremely contractile. It can lengthen and contract, producing peristaltic movements that facilitate the ingestion of food.

11.2. Middle or gastric region

It contains the part of the digestive tract that is usually covered by large cecum. In this region is the resophagus, which is a rectilinear portion of the digestive tract. It joins the stomach to the base of the tube, and has a fairly narrow lumen in the centre, which is frequently star-shaped due to the presence of longitudinal folds. The middle portion of the digestive tract is a flattened, vacuolated cavity with a pair of diverticula at the first ring of each traversed somite.

The stomach has a powerful musculature made up of branched fibres. The branches of the different fibres intersect obliquely, forming a tightly woven network.

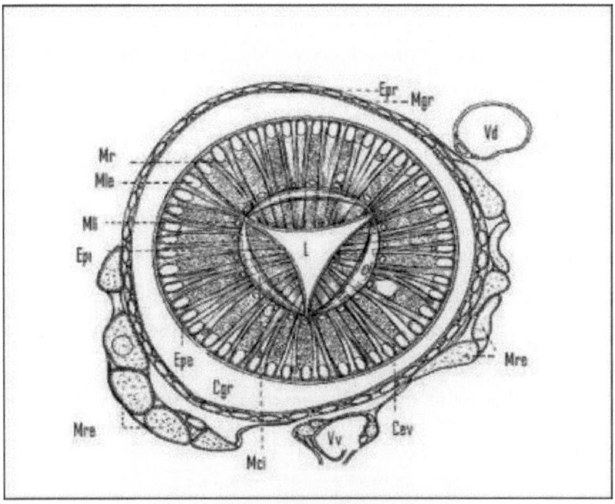

Fig. 6 : Cross-section of the proboscis.
Cev: salivary gland evacuating canaliculi in *Haementeria costata*, Cgr: rostral sheath cavity, Epe: rostral external epithelium, Epi: rostral internal epithelium, Epr: rostral sheath epithelium, L: rostral cavity, Mci: circular muscles, Mgr: musculature of the rostral sheath, Mle and Mli: external and internal longitudinal musculature, Mr: radial muscles, Mre: retractile muscles of the rostrum, vd: dorsal vessel; vv: ventral vessel (Grasse, 1959).

11.3. Terminal or intestinal region

It is shorter than the previous region and usually bears small cecum. A narrowed portion, fitted with a sphincter, forms a true pylorus and forms the transition between the gastric region and the intestine. The intestinal tract carries four pairs of diverticula, which are larger than the gastric cecum and are located above the testicles. The epithelial cells here are less elevated than in the stomach, but their striated cuticle is much more pronounced and much higher. In many species, the intestine has no musculature of its own. It is sometimes provided with a few fibres of the lateral type. This portion ends in a rectal canal preceded by a rectal ampulla and leads to the dorsal anus.

12. Reproduction

Leeches are hermaphrodites, with paired gonads that are generally in a fixed position. There is always a single pair of ovaries situated in front of several pairs of testes. There is usually only one orifice for each sex. The male orifice is always located in front of the female orifice. These orifices open in the clitoral region on somites X and XI.

12.1. Male device

The male genital tract (Fig. 7) consists of several pairs of testicular sacs, the number of which varies from species to species. These sacs are arranged segmentally from somite XII, alternating with the cecum of the stomach. Each testicle has an efferent duct which leads to the deferent duct which, on either side,

ascends to the eleventh somite, passing beneath the digestive cecum. Each ductus deferens then widens into a seminiferous vesicle, then narrows again to become an ejaculatory duct, the terminal part of which is filled with glands **(Minelli, 1979)**. The two ejaculatory ducts unite at the front in a common chamber or atrium. The atrium develops the spermatophore in indirectly fertilised species, whereas it becomes a functional copulatory organ in directly fertilised species. The male orifice is located between the second and third rings of the eleventh somite.
In the Erpobdellidae, there is a multiplication in the number of testes, with loss of the segmental device. Disseminated throughout the parenchyma, they are linked by narrow efferent ducts to the common deferent duct, which is also entirely embedded in the parenchymal tissue.
In the Hirudinidae, the male apparatus is more complex, with 9 to 10 pairs of testicles.

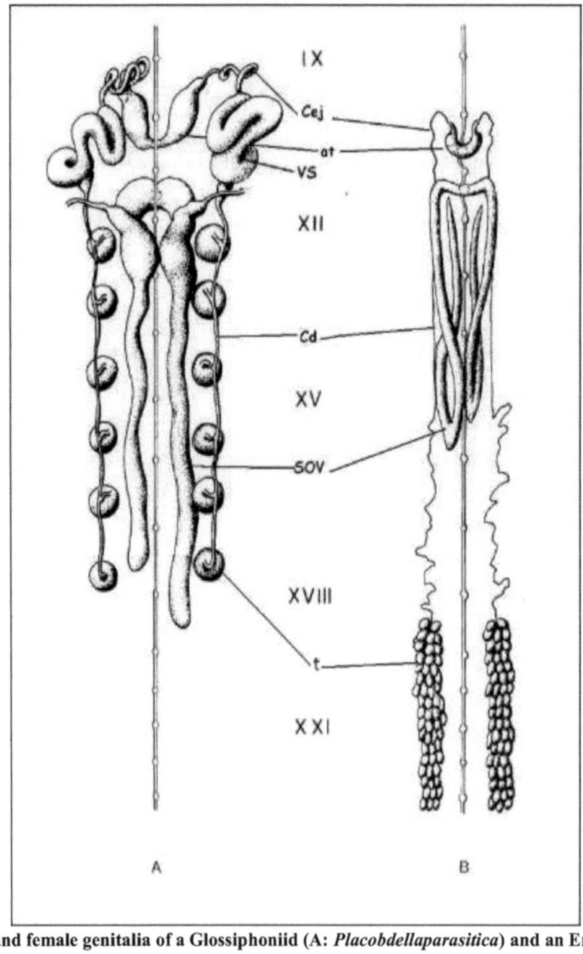

Fig. 7 Male and female genitalia of a Glossiphoniid (A: *Placobdellaparasitica*) and an Erpobdellid (B:

***Erpobdella sp.*) The Roman numbers correspond to the metamerism. ca: caecum of the ovarian sac, at: atrium, ce: efferent duct, cej: ejaculatory ducts, cd: deferent duct, vs: seminal vesicle, sov: ovarian sac, t: testes (Minelli, 1979).**

12.2. Female device

The female genitalia consist of 2 elongated sacs or ovisacs (Fig.7) placed in the ventral median lacuna, next to the ejaculatory ducts, below the digestive tract. During the sexual resting period, they are just two small cylindrical cords occupying the length of 2 segments. During the active period, they grow considerably. The ovarian sacs unite at the female orifice to form a common oviduct.

Each ovisac can be considered as a portion of a creloma bounded by a well-defined wall and containing the ovary, which is a solid cellular cord.

13. Coupling

Mating takes place in two distinct ways:

- **Indirect mode**

By implantation of a spermatophore in the wall of the partner, this mode is characteristic of species without a penis (Rhynchobelliformes (have a tube) and Pharyngobdelliformes (have a pharynx without a jaw). Fertilisation is hypodermic.

- **Direct mode**

By intromission of the penis of one partner into the vagina of the other. This method is characteristic of the Gnathobdelliformes (which have a pharynx armed with a jaw). Fertilisation is direct: the male copulatory organ of one partner penetrates the interior of the female gonopore of the other partner, in this case fertilisation is reciprocal **(Kutschera et Writz, 2001)**. It should be noted that all leeches are oviparous and their oviposition, formed by a certain number of eggs characteristic of the species, is enclosed in a cocoon.

1. .1. Rhynchobdelliformes

Mating involves the emission of a spermatophore whose surrounding mucus hardens on contact with water. Each spermatophore consists of 2 tubules which are open at their proximal end. They are filled with sperm and their proximal parts contain a granular plasma with proteolytic properties **(Doin, 1952)**.

The spermatophore is expelled and then attached to the partner's wall by the contraction of the atrium muscles. It empties its sperm into the tissue of the recipient, whose teguments are lysed by a protease contained in the spermatophore. Injection of the sperm is due to pressure produced by the tension in the walls, which retract on contact with water. Glossiphoniid spouses, taken as an example from the Rhynchobdelliformes, are held together by their mouth suction cup, which they attach to a point on the clitellum. The spermatophores are then deposited in this area, followed by injection of the sperm into the general cavity, infiltrating the vector tissue from one end to the other (Fig. 8). Self-fertilisation has been observed in Glossiphoniidae by *Helobdella robusta* **(Weisblat & Huang, 2001)**.

2. .2 Pharyngobdelliformes

In the Erpobdellidae, taken as an example from the Pharyngobdelliformes, the spermatophores consist of a thick basal plate surmounted by two conical extensions. Their walls are so thick and rigid that the sperm is only injected through the tissues thanks to the powerful musculature of the ejaculatory ducts. The spermatophore does not leave the sexual organs during copulation.

13. 3. Gnathobdelliformes

The Hirudinidae, taken as an example of the Gnathobdelliformes, move towards each other belly to belly and in opposite directions so that the male orifice of one is superimposed on the female orifice of the other and vice versa.

Copulation occurs by intromission of the penis into the female genital tract. In this case there are no spermatophores and therefore no hypodermic copulation. The species of this order are all characterised by the presence of a penis.

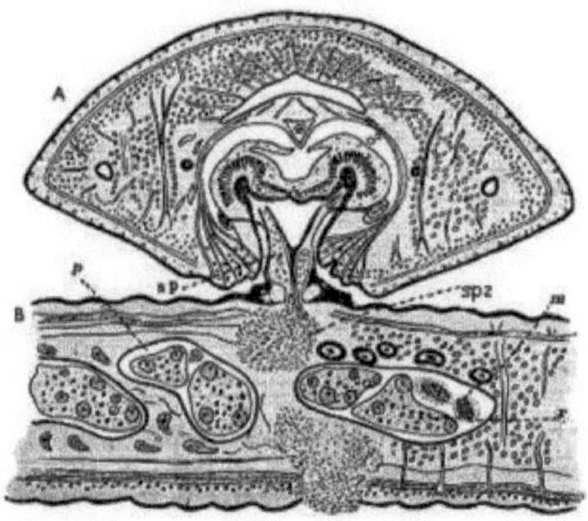

Fig. 8 : Section through the mating point of two *Erpobdella octoculata*.
m: mass of sperm entering the ovary, p: ovarian wall infested with sperm, r: ovarian rachis, sp: spermatophores of subject A (CT donor) injecting sperm into the tissues of subject B (CL recipient), spz: sperm (Grasse, 1959).

14. Eggs and cocoons

There is a mucous hypersecretion of the clitellar region at the time of egg-laying. This region forms a sort of belt for the animal, which is freed by repeated contractions of the whole body.

Fig. 9 : Cocoon of *Glossiphonia complanata* attached to a support.
The eggs are clearly visible through the transparent wall of the cocoon (Grasse, 1959).

The second eggs are placed in a cocoon. The cocoon is a sort of ovoid capsule of variable consistency, bounded by a substance that appears horny (Fig. 9).
Oviposition occurs at variable times after mating: (*Glossiphonia complanata:* 8-21 days, *Hirudo medicinalis:* 1-19 days, *Erpobdella sp.*: 2-20 days).
Spawning periods are in the northern hemisphere from March to October, depending on the species **(Grasse, 1959).**
The young are fed by a protein-rich liquid secreted by the clitelline glands and contained in the cocoon.

14.1. Glossiphoniidae

Their cocoons are tough, horny and translucent. They are glued to supports or held to the belly of the animal.
Glossiphonia complanata lays 3-8 transparent cocoons each containing 11-50 eggs once a year.
Glossiphonia heteroclita lays eggs 2 to 3 times a year, each time producing a cocoon with around thirty eggs.

14.2. Piscicolidae

In this case, the cocoons are applied against various supports: stones, plants, etc. One fish farm is capable of producing 50 cocoons. These have a chitinous-looking envelope and are full of an albuminous substance secreted by the clitellate glands. Each cocoon contains only one egg.

14.3. Erpobdellidae

In *Erpobdella octoculata*, cocoon development lasts around twenty minutes. It is scale-like, flattened and attached to immersed plants. Each cocoon encloses several eggs.

14.4. Hirudinidae

Hirudo medicinalis cocoons are fairly firm in consistency and are 2 to 3cm long and 1cm to 1.5cm wide. They are placed on land even when the leech is aquatic. Under the influence of humidity, their envelope becomes spongy and a large number of small eggs can be seen inside.

15. Young people's relationship with their mother

Glossiphoniidae do not abandon their progeny. They cover their cocoons with their bodies and, when they hatch, the young attach themselves to the ventral face of their mothers with their posterior suckers. *Marsupiobdella africana* also has an incubation chamber.

16. Reproductive cycle

- *Glossiphonia complanata* and *Erpobdella octoculata* breed at the age of one year and a second time at the age of 2 years; each brood of 26 young. **Mann (1957)** suggests that some *Glossiphonia complanata* can live for 4 years.
- *Helobdella stagnalis* breeds in spring. After the young are released, the breeders die in May. Their offspring may breed in the autumn, but part of the population overwinters and breeds in the spring of the following year. Longevity is less than 12 months.
- *Theromyzon tessulatum* lays from 60 to over 600 eggs. It breeds only once and then dies. Its lifespan is estimated at 17 months.

III. ECOLOGY

Ecological observations were made on freshwater leeches. Freshwater is classified according to its organic matter content into :

J Eutrophic :

A watercourse that has become enriched with dissolved nutrients, visually evident by abundant aquatic plant growth.

J Oligotrophic :

A watercourse characterised by low biomass and low availability of nutrients, trace metals and/or growth factors.

J Dystrophic :

These waters are characterised by major chemical changes, due to excessive nutrient inputs accompanied by permanently low oxygen levels or anoxia, leading to the elimination of most exploitable benthic and demersal fish stocks.

In Palearctic regions, very dystrophic waters generally contain no leeches, while oligotrophic waters are regularly frequented by *Erpobdella octoculata* and eutrophic waters by *Helobdella.*

The pH of the water influences the distribution of several species. In fact, the majority seem to prefer neutral or slightly alkaline water, while others can cope with low pH levels. In general, however, leeches cannot tolerate high acidity and inhabit water with a pH of between 6.3 and 7.2 **(Minelli, 1977).**

Temperature in turn appears to be an important factor in reproductive activity in leeches. Indeed, **Biernacka and Davies (1994)** have shown that in *Nephelopis obscura,* temperature and food intake have an influence on gonad biomass, as does *Helobdella triserialis* **(Gullo, 2003)**. For each species, there is a temperature below which reproduction is absent (16° for *Hemiclepsis marginata*, 12° for *Theromyzon* *tessulatum*) **(Perret, 1952).** On the other hand, letal temperature seems to oscillate between 33°c and 40°c **(Minelli, 1977)**. Temperature does not play a major role in the colour change process **(Sawyer et Davies, 1974).**

IV. CLASSIFICATION

The classification of the Hirudinees varies according to the authors; in this overview we present the most recent proposed by **Govedich (2001)** and the most classic proposed by **Grasse (1959)**.

1. Classification by Govedich, 2001

Govedich (2001) proposed the following classification:

Phylum : Annelida

Class: Clitellata Subclass: Hirudinea Subclass: Acanthobdellida Subclass: Euhirudinea

Order: Rhynchobdellida

Families: Glossiphoniidae

Piscicolidae

Ozobranchidae

Heptacyclidae

Platybdellidae Mysiodobdellidae Austrobdellidae

Order : Arhynchobdellida

Superfamily: Erpobdelloidea

Family: Erpobdellidae

Superfamily: Gnathobdelloidae

Family: Ornithobdellidae

Limnobdellidae

Hirudinidae

Haemadipsidae

According to **Govedich (2001),** the Hirudinees include the Euhirudinees and the Acanthobdellidees. The latter have affinities and characteristics in common with the Oligochetes (presence of silk), while the Euhirudinees include all the "true leeches".

The Euhirudinees are divided into two orders:

- Rhynchobdelles: They have proboscis.
- Arhynchobdelles: They have no proboscis.

This order comprises two superfamilies:

- Erpobdelloidea: These are leeches with pharyngeal stylets.
- Gnathobdelloidae: These are leeches armed with jaws, the number of which plays an important role in classification. In fact, there may be two jaws (Dignathes) or three (Trignathes). The jaws are equipped with a number of teeth characteristic of the species, arranged in a single row (Monostichodontes) or in two rows (Distichodontes).

2. Grasse classification, 1959

According to **Grasse, 1959** the Hirudinees are divided into 4 orders, 127 genera and more than 300 species.

2.1. Order Acanthobdelliformes

This order is remarkable for the characteristics it has in common with the Oligochetes. It contains only one family with a single genus.

J Family Acanthobdellidae

The single genus *Acanthobdella* forms a transitional group with the Oligochetes; it has a single posterior sucker. Segments 2 to 6 bear 4 double rows of setae. The coelomic cavities are recognisable. These animals are parasites of Salmonids.

2.2. Order Rhynchobdelliformes

This order is characterised by the presence of a devaginable proboscis and a developed vascular system. There are no terrestrial species.

Family Glossiphoniidae

Species in this family have a dorsoventrally flattened foliaceous body. The internal organs are visible through the skin. The anterior sucker is only slightly developed compared with the posterior sucker, which is discoid. The jaws are absent and replaced by a proboscis. The number of eyes varies from species to species. Glossiphonidae are often parasites of freshwater invertebrates such as *Glossiphonia* (clepsine) and live off physes (molluscs). They are not good swimmers and are usually found clinging to the surface of the water.

substrate. This family is characterised by a high level of parental protection. The young attach themselves to the mother's belly, which feeds them with an albuminous secretion.

Piscicolidae family

Members of this family have a cylindrical body with highly-developed suction cups. They are parasites of fish: *Pontobdella* parasites rays, *Branchellion* torpedoes and *Piscicola* freshwater fish.

2.3. Order Pharyngobdelliformes

Family Erpobdellidae

This family is characterised by a vermiform cylindrical body and a mouth that occupies the whole of the front sucker. Pharyngeal stylets may be present. The jaws and digestive caeca are absent. Species in this family are predators. They feed on insect larvae and Oligochetes. The scale-like cocoons are found attached to stones or submerged plants.

Family Semiscolecidae

Semiscolecidae have a jaw and rudimentary gastric caeca. There are usually more than one pair of testes per segment.

J Family Trematobdellidae

Species in this family have a digestive tract with a narrow valve that allows it to communicate with the outside world.

J Americobdellidae family

These are neotropical Hirudinees, in which the middle somite comprises 5 rings.

There are six pairs of eyes and nine pairs of testes. The genital pores are separated by 5 rings. The penis and vagina are absent. The saccate ovaries lead to the female gonopore. There are 17 pairs of nephridia. The mouth and pharynx are large. The jaws are absent.

Family Xerobdellidae

They are terrestrial, hematophagous hirudines. The genus *Xerobdella* has 8 pairs of eyes and a pair of tentacular appendages on the fifth segment. The genus *Diestecostoma* has a decamere somite.

2.4. Order Gnathobdelliformes

Family Hirudinidae

The Hirudinidae have a cylindrical, vermiform body, three pairs of jaws armed with 40-60 well-developed teeth and five arched eyes. Species in this family are considered good swimmers. They are typically freshwater leeches, hematophagous ectoparasites of mammals, birds, reptiles and amphibians. This family has no parental protection for juveniles.

Family Haemadipsidae

These land leeches live on shrubs and trees. *Haemadipsa* attacks humans and livestock and is a veritable scourge for travellers.

Members of this family have a pharynx that is not very muscular, long and without a jaw, and a digestive tract without caeca. The testicles are small, numerous and scattered.

Later, other classification studies were carried out, including the work of **Soos (1966, 1967, 1968** and **1969**) and that of **Sawyer (1986)** and **Govedich (2001),** an overview of which has already been given. In our systematic study, we used the determination keys proposed by Sawyer (1986). This support is considered to be the most complete and detailed.

V . PARASITISM

In addition to the spoliating action that certain Hirudines exert on the human organism, they are capable of transmitting various human parasites. These parasites are present in their bodies and are transmitted hereditarily.

Hirudo medicinalis has been studied as a possible germ vector. This leech can store in its stomach koch's bacillus (60 days), Eberth's bacillus (7 days), paratyphic bacillus B (3 months), pig rouget bacillus (10 days), anthrax bacterium (10-14 days), exanthematic typhus virus (2 days), swine fever virus (20 days), avian smallpox virus, recurrent fever spirochete (4-20 days), avian spirochetosis (7 days), *trypanosoma brucei* (3 days) and *T. equiperdum* (4 days), and malaria parasites (3-4 days).

These live parasites are readily available, not only for purely scientific purposes, but also for therapeutic purposes (pyretotherapy, malaria therapy, general paralysis therapy, etc.) **(Grasse, 1959).**

VI . THERAPY

Leeches (*Hirudo medicinalis*) have been used in medicine since ancient times. The image of the leech is that of a slimy worm that feeds on blood and can transmit diseases to humans.

Hirudin, extracted from the saliva of leeches, has countless virtues. In fact, it is analgesic, anticoagulant, antibacterial, anti-inflammatory and anaesthetic. All these characteristics are of interest in the treatment of arthrosis (deterioration of joint cartilage, causing chronic pain, stiffness, swelling of the joints and sometimes the inability to move). The great Louis Pasteur applied it to his face to alleviate the symptoms of his facial paralysis **(Grasse, 1959).**

Our present work is a preliminary study of the biodiversity of leeches in Tunisia. It has enabled us to obtain the following results:

CHAPTER II

MATERIALS AND METHODS

1. MATERIAL

1. Presentation of the study environment

1.1. Geographical location

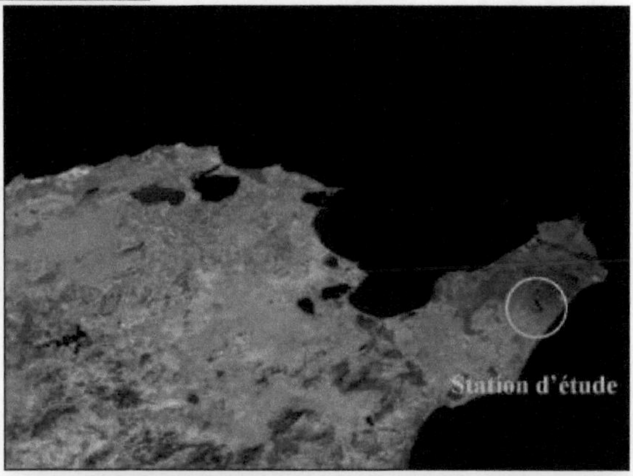

Fig. 10: Map of the Gulf of Tunis (N.A.S.A, 2006).

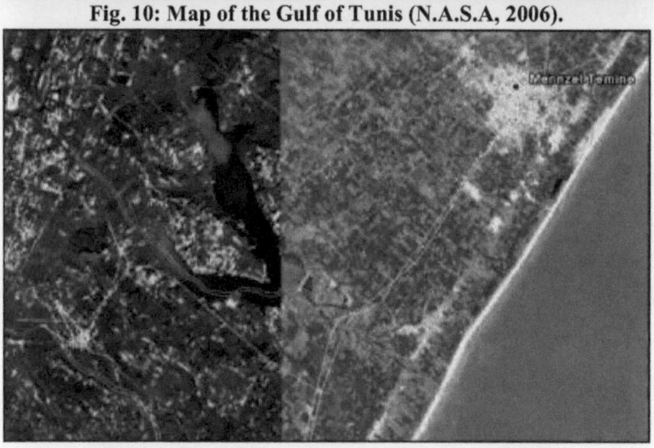

Fig. 11: Map of the Lebna dam (N.A.S.A, 2006).

The Lebna dam is located in Cap Bon, around 80 km from the city of Tunis, near the town of Menzel Temime in the governorate of Nabeul. It holds back the waters of the Oued Lebna. The latter has two tributaries: oued El Oudiane and oued Bou Dokhane at Henchir El Bendara, linking the Korba region to Menzel Temime before flowing into the Mediterranean Sea. The oued Lebna is characterised by stagnant water with a basic affinity, the salinity being between 0.57g /1 and 2.9g/l, **(Harrath, 1999).**

Fig. 12: Photo of the Lebna dam.

The samples of Annelides Achetes were collected from the dam installed on the course of the Oued Lebna. The latter is polluted by waste from the tomato factory next door.

1.2. Withdrawals

Sampling was carried out at the edge of the 22 m long, 0.5 m wide curb, covering an area of 11 m^2 at the level of the intake tower (Fig. 12).

Monthly harvests were carried out for ecological and biometric studies.

2. Materiel Biologique

2.1. Capture

Hirudinees are lucifuges. Individuals can be found under pebbles. They are delicately collected using a moistened brush (Fig. 13), then placed in a jar containing water from the dam. During certain periods of the year, when the temperature is quite high (around 30°C), we have noticed that the animals transported cannot resist and perish during transport. To avoid this inconvenience, it is advisable to transport them in temperature conditions comparable to those of their biotope by placing them in an isothermal environment such as an icebox.

Fig. 13: Collecting Hirudinees in the field.

2.2. Breeding in the laboratory

The leeches collected can be reared in the laboratory for several months. They are placed in sealed glass bottles, as they may emerge and die. The bottles must be aerated from time to time. These worms are fed at least once a week with small pieces of liver, which are given and removed for a day. The water used for rearing the worms comes from the dam where the worms were harvested. It is changed daily or every other day, after removing the waste, of which mucous secretions make up the majority.

We were, however, able to breed in tap water collected 2 to 3 days in advance, in another container in order to minimise the effect of the bleach.

II. METHODS

This involves taking samples of ecological and environmental parameters and studying the animals in the laboratory.

1. Ecological parameters

1.1. Abiotic factors

a. The temperature

Every month, the temperature of the water is measured a few centimetres below the surface using an ordinary thermometer graduated in degrees Celsius.

b. Salinite

Salinity is measured using a YSL model 33 salinometer. Measurements are taken monthly during the study period.

c. The pH

The pH of the water is determined using a GENWAY model 3510 ph meter in the laboratory.

In addition to the abiotic factors of the environment, the biotic factors make it possible to treat the stand where the species lives.

c.2. Calculation of some parameters

a. Specific richness (S)

The specific richness of a stand is the total number of species found in a system over the period of one year.

b. Relative abundance (Ar)

The relative abundance of a species is defined by the ratio of the number of individuals of that species **(n)** to the total number of individuals harvested **(N)**; this ratio is reduced to 100 :

$$Ar = n / N \times 100$$

This percentage reflects the importance and dominance of the species in the stand.

c. Average density (Dm)

It corresponds to the ratio of the total number of individuals (n) to the total surface area of the samples (Sp = $11m^2$). It is expressed as the number of individuals per m^2 :

$$Dm = n / Sp$$

2. Techniques for studying animals in the laboratory

2.1. External and internal morphology

The different species are first sorted using the naked eye. Morphological characteristics such as the number of eyes and their location, the number of rings separating the two gonopores (male and female), the digestive system including the number of digestive caeca and their structure (single or lobe) and the reproductive system including the number and structure of testis and ovisac, were examined. Laboratory observation was carried out using a stereoscopic magnifier, on specimens fixed in 70% alcohol of various ages and sizes, as well as on dissected specimens. Some measurements were taken using a micrometer incorporated into the eyepiece of a magnifying glass. The representative criteria are photographs for the descriptive study.

2. 2. Biometrics

The biometric study is carried out in order to study the sexual cycles. The animals collected and brought back to the laboratory are directly sorted and classified by species. They are then placed one by one in a crystalliser, with millimetre paper placed underneath. We wait until the animal has reached its maximum extension, and then measure its length in centimetres. We tried to anaesthetise the animals but found that the worms contract under the effect of alcohol.

2. 3. Data processing

The results obtained are expressed as averages followed by the standard error of the mean. The latter is represented on the graphs by error bars.

Calculation of the average (m)

The mean (**m**) is equal to the sum of the data, divided by the number of data:

$$m = \frac{\sum_{i=1}^{N} X_i}{N}$$

x_i = different values of the variable.

N = sample size.

Calculation of the standard error (Sm)

$$Sm = \frac{S}{\sqrt{N}}$$ **(Schwartz, 1963)**

N = sample size

S = the standard deviation given by the following relationship :

$$S = \sqrt{\frac{\sum (xi - m)^2}{N - 1}}$$ **(Schwartz, 1963)**

Statistical analysis

Correlations between the numbers of each species and the three ecological parameters (T, pH and Salinity) were studied using Microsoft Excel and STATISTICA 6.0 software (Spearmann test). A Spearmann coefficient R was calculated and a probability value p less than or equal to 0.05 was considered statistically significant.

2. 4. dissection

In order to see the absence or presence of the proboscis and to note the anatomical differences of the genitalia in some species of these Hirudinae, a dissection must be made.

Mature, large specimens are selected from the collected animals for dissection. The preferred method for sacrificing these animals is to place them for a few minutes inside a container containing 70° alcohol. The animal is then fixed dorsally with two pins, inserted into the two suction cups. Note that for the front suction cup, the pin must be inserted at the level of the upper lip so that the jaws can be observed.

Using forceps and a scalpel, the skin stuck to the musculature is lifted along a longitudinal line.

2.5 Histological techniques (Martoja and Martoja, 1967)

The aim of the histological technique is to reveal the presence or absence of the proboscis in some species of the Hirudineae, and the digestive caeca in *Erpobdella testacea.*

This technique, which produces permanent preparations that are perfectly observable under the microscope, involves four successive operations:

J **Mounting**

The aim of this operation is to preserve the cells in a state as close as possible to the living state. However, there is no such thing as a universal fixative. All mixtures have advantages and disadvantages. As a result, it is necessary to choose those that give the best results, depending on the goal being pursued. In this work, the fixative chosen is Bouin Hollande, which can be preserved for a long time without damaging the tissue.

Bouin Hollande

- Distilled water100ml
- Neutral copper acetate2 .5g
- Picric acid4g
- Formol 40%10ml
- Acetic acid, crystallizable1 .5 ml

Before being fixed, the animals intended for histological study are anaesthetised with pre-heated water (30° to 40°), to avoid sudden contractions of the body during contact with the fixative. Such movements can lead to deformation and even fragmentation of the animal. The animal is then immersed in vials containing the liquid fixative. The animal remains in the Bouin for at least 3 to 4 days. The animal is then washed in tap water for 24 hours to remove any excess fixative.

J **Inclusion**

The aim of this operation is to penetrate the pieces with a substance of a favourable consistency for cutting and holding the cells in place. There are several inclusion media, the most common of which is paraffin. Celloidin or gelatin inclusions are of much more limited use. Paraffin is a hydrocarbon that cannot be mixed with water or alcohol. For this reason, prior treatments are necessary before the actual embedding. These involve dehydration, penetration by an intermediate liquid and impregnation by molten paraffin before final blocking.

❖ Dehydration

Paraffin is soluble in xylene, toluene, butyl alcohol and dioxane. It is therefore necessary to carefully dehydrate the pieces before including them.

The most common desiccant is ethyl alcohol. The pieces are successively placed in baths of increasing concentration:

o 3 baths in 70° alcohol, each lasting 20 minutes

o 3 baths of 95° alcohol, each lasting 20 minutes

o 3 baths in 100° alcohol for 20 minutes each

❖ Penetration by an intermediate liquid

Given the insolubility of paraffin in ethyl alcohol, it is necessary to replace the latter with an intermediate liquid which must be miscible with the alcohol and the paraffin. Butanol or butyl alcohol, which is also a stand-by liquid, is the most commonly used.

The operation involves replacing the 100° alcohol with the intermediate liquid. The success of inclusion will depend on the success of this operation.

o Butanol (1)1 hour

o Butanol (2)Inuit

o Butanol (3)1 hour

o

❖ Penetration with paraffin

After passing through the intermediate liquid, the liquid paraffin is very carefully impregnated. This impregnation is carried out hot, in a controlled oven at the melting point temperature of the paraffin chosen or at a very slightly higher temperature (58°). In this case too, it is advisable to use three successive baths in order to completely eliminate the intermediate solvent.

o Paraffin(l) 1 hour

o Paraffin(2) Inuit

o Paraffin(3) 4 hours

❖ Blocking

Once the hot penetration of the paraffin is complete, the organs are placed in blocks and can be stored indefinitely. This operation consists of pouring melted and previously filtered paraffin into plastic moulds, and quickly placing the part in them, orientating it appropriately with lightly heated tongs.

Before the paraffin cools, which takes about ten minutes, a reference label is placed in each block, which also serves to identify the cutting plane. When the paraffin has cooled completely, it is demoulded to obtain solid blocks of paraffin, containing the part ready to be cut into thin sections by the microtome.

***J* Cutting and gluing**

For best observation, cuts should have an average thickness of between 5 and 7 ^m.

❖ Microtomy

Before the cuts are made, the blocks must be trimmed around the workpiece to give the cross-sectional area a trapezoidal shape.

The block, which is cut and glued to the previously heated object holder, is placed on the base of a microtome, fitted with a razor and a mechanism for advancing its base, enabling the block to be sliced to the required thickness.

Due to the heat produced by friction with the razor, the paraffin melts slightly, allowing the cuts to adhere to each other and form a ribbon.

The two edges of the section to be cut (top and bottom) must be exactly parallel to obtain a straight ribbon of successive cuts of variable length.

❖ Gluing cuts

The series of sections must be placed on a glass slide (cover slip) and firmly adhered for subsequent handling. After splitting the strips into pieces according to the length of the coverslips, the sections are spread out and glued to the albumen water prepared in the following way:

- Prepare a mixture of egg white and glycerine in equal parts in a container containing glass beads.
- Add 1g sodium salicylate to prevent mould growth in the medium.
- Filter through water-moistened filter paper (filtration takes around two days).

Everything is kept in the fridge. Gluing the cuts involves the following steps:

- Each blade must be numbered with a diamond in a corner that is always the same. This is to maintain the serial number of the cuts on the blade in question.
- Using a small glass stirring rod, place a small drop of glycerol albumin on the slide and spread it with the pad of the index finger.
- Place the slide on a temperature-controlled heating plate (4 to 5°C below the melting point of the paraffin) and pour a thin layer of water over it.
- The cuts are placed on the water, shiny side up against the blade: they spread out.

and they can be unfolded; needles or tweezers are used to help unfold them.

- After spreading, the excess liquid is removed using filter paper.
- Finally, the slides are left to dry in a dust-free environment, either at room temperature or in an oven at around 40°C.

***J* Colouring**

Sections simply mounted on slides are insufficient for microscopic study, since the only way to distinguish the structure of a tissue is by the difference between the refractive indices of its components. However, the cellular components all have roughly the same optical density.

However, it is possible to stain the tissue and reveal its microanatomical make-up due to differences in the chemical composition of the structures, which react differently to the dyes, giving contrasts on the section.

Eosin and Toluidine blue are used as stains: nuclei, basophilic cytoplasms and certain secretions are stained violet blue, muscle elements orange and acidophilic cytoplasms pink.

The first step is to remove the paraffin using a solvent, usually toluene. This is called dewaxing and is followed by hydration, as the dyes are generally found in aqueous solutions.

❖ Dewaxing

The slides are placed on the stage set at a temperature 1 to 2° above the melting point of the paraffin. They are then passed through the following baths:

o Toluene (1) 5mn

o Toluene (2) 5mn

❖ Hydration

This is done by passing the blades through alcohols of different concentrations.

decreasing : o Alcohol 100°5mn

- o Alcohol 95°5mn
- o Alcohol 70°5mn
- o Water (1)5mn
- o Water (2)5mn

❖ Toluidine eosin stain

o Toluidine blue 0. 5%3mn

o Rinse with running water

o 1% aqueous eosin15seconds

o Rinse with running water

❖ Dehydration

The sections are dehydrated again by soaking in two baths of absolute alcohols and then passed through a lightening agent, usually Toluene.

o Alcohol 100° (1) 1mn

- o Alcohol 100° (2) 2mn
- o Toluene (1) 5mn
- o Toluene(2) 5mn

❖ Cutting assembly

■ Apply a drop of Baume de Canada that is neither too thick nor too fluid.

■ Present a cover slip, previously soaked in toluene, tilting it gradually so that the Canada Balsam gradually spreads and covers the cut without trapping air bubbles.

After the coverslip has been dried at room temperature or in an incubator (35 to 40°C), the preparations are ready for observation under the microscope. Representative sections are photographed.

CHAPTER III

SYSTEMATICS: MORPHOLOGICAL AND ANATOMICAL STUDY

In view of the absence of determination keys for North Africa, which is part of the Palearctic region, in this systematic work we have adopted the criteria used by **Sawyer (1986)** and Nesemann and Neubert (1999) for Europe and the Palearctic region. For the determination of *Hirudo troctina* we used the criteria of **Hetchel and Sawyer (2002).**

I. KEYS FOR DETERMINING SPECIES IN EUROPE AND THE PALEARCTIC REGION (SAWYER, 1986)

Glossiphoniidae

1 - Four pairs of eyes (*Theromyzon)* 2

- One, two or three pairs of eyes3

2 - Gonopores separated by two rings *Theromyzon maculosum*

' Gonopores separated by four rings *Theromyzon tessulatum*

3 - A single pair of eyes (non composd) 4

- Two or three pairs of eyes (sometimes appearing as a single pair) 5

4 - Brown plate in the dorsal region of the+ *Helobdella stagnalis*

- Absence of brown patch in the mid-dorsal region of the neck *Batracobdella algira*

5 - Antdorsal sucker not distinct from the body: *Hemiclepsis marginata* ' Antdorsal sucker not distinct from the body6

6 - Pore of the mouth near the front edge of the front sucker *Placobdella nabeulensis*

Papillae on the dorsal surface, seven pairs of digestive caeca - Pore of the mouth in the middle of the anterior sucker *7*

7 - Gonopores with a common pore *Alboglossiphoniahyalina*

Several infra-specific varieties have been recognised: var. *hyalina, striata* and *papillosa.*

' Gonopores sdpards by two rings8

8 - Six pairs of digestive caeca9

- Seven pairs of digestive caeca10

9 - Two black-brown dorsal paramedian lines regularly interrupted by papillae *Glossiphonia sexoculata*

- Two brown dorsal lines irregularly interrupted by papillae *Glossiphonia concolor*

10 - Dorsal surface with tubercles on the second and third rings of each middle segment *Glossiphoniaverrucata*

- Smooth dorsal side11

11 - Six pairs of testis *Glossiphoniapaludosa*

- Ten pairs of testis *Glossiphoniaslovaca*

Hirudinidae

1 - Five pairs of eyes, third and fourth pairs separated by a single ring4

- Four pairs of eyes (rarely three pairs), third and fourth pairs separate by two rings (*Xerobdella*) 2

2 - Gonopores separated by three - three and a half rings *Xerobdella lecomtei*

- Gonopores separated by four or more rings3

3 - Gonopores separated by four and a half rings (four -five) *Xerobdella anulata*

- Gonopores separated by six and a half rings *Xerobdella praealpina*

4 - Jaws with two rows of twenty or more teeth each (Distichodontia) *Haemopissanguisuga*

- Jaws with a single row of thirty or more teeth (Monostichodontia) 5

5 - Jaws with salivary papillae *Limnatisnilotica*

- Jaws without salivary papillae *Hirudo medicinalis*

Longitudinal fissure at the level of the upper lip absent; each macbory has between 60 - 100 teeth; relative form: *Hirudo* troctina characterised by the presence of a zigzag outline on the ventral face of the animal.

Erpobdellidae

1 - Segments of the mid-body shaped mainly by five rings2

Three to four pairs of eyes are present

- Midbody segments not shaped by five rings *5*

2 'Conduit djaculateur sans une boucle preartiale 3

- Ejector pipe with a preartial loop 4

3 - Gonopores separated by two and a half rings *Erpobdella octoculata*

- Gonopores separated by four rings *Erpobdella testacea*

According to Hetchel and Sawyer (2002):

1 - Dorsal side with four or six metameric black spots *Hirudo troctina*

Right vagina, not wrinkled, bulbous, massive epididymides in relation to the ejaculatory bulb.

2 - Dorsal surface predominantly longitudinally striped, with metameric black spots.

absent *Hirudomedicinalis* .

Pleated vagina, dpididymis and ejaculatory bulb of almost equal width.

11. INVENTORY OF SPECIES COLLECTED

The preliminary systematic study of the Annelidae Achetes of Tunisia has enabled us to identify with certainty 6 species belonging to 6 genera and three different families in the Lebna dam, based on morphological analysis, oviposition type and diet (Table 1). Two other specimens that can be identified as two other distinct species are designated in the table below as follows: *Limnatis sp.1* and *Limnatis. sp.2.*

Table 1: List of species encountered at the study site.

Branch	Class	Sub-class	Order	Family	Genus and species
Annelides	Clitellates	Hirudinees	Rhynchobdelles	Glossiphoniidae	*Batracobdella algira*
					Theromyzon tessulatum
					Placobdella nabeulensis
					Alboglossiphonia hyalina
			Arhynchobdelles	Hirudinidae	*Hirudo troctina*
					Lilnatis nilotica
					Limnatis sp.l
					Limnatis sp.2
				Erpobdellidae	*Erpobdella testacea*

111. DESCRIPTION AND IDENTIFICATION OF SPECIES

1. Family Glossiphoniidae Vaillant, 1890

The Glossiphoniidae family is considered to be the richest in terms of described species and the most diverse of the freshwater Hirudinae. It is found on every continent except Antarctica **(Figueroa, 2006).** This group mainly comprises

Hirudinae with dorso-ventrally flattened bodies, which normally feed on turtles and amphibians. It is the only family in the Achetes class that displays a high level of parental protective behaviour **(Paez *et al.*, 2004).** Moreover, **Kutschera (1986)** noted that *Helobdella stagnalis* captures insects and Oligochetes, mainly *Tubifex*, to feed its juveniles attached to its belly.

On the basis of the number of eyes and their position, the number of rings separating the male and female gonopores, the number of gastric caeca and the number of testes, we were able to identify four distinct species: *Batracobdella algira* , *Theromyzon tessulatum* , *Placobdella costata* and *Alboglossiphonia heteroclita.*

1.1.*Batracobdella algira* (Moquin-Tandon, 1846)

Glossiphonia algira **Moquin -Tandon, 1846** (I .ocalite typique: Algerie: se;tif)

Clepsine algerica **Diesing ,1850**

Batracobdella latasti **Viguier ,1879**

Glossosiphonia algira **Wolterstroff, 1900**

Batrachobdella algira **Johansson, 1926**

Placobdella schottlaenderi **Johansson, 1926**

Batracobdella algira **Autrum, 1936**

Like all species of the Glossiphoniidae family, *Batracobdella algira* has a dorso-ventrally flattened body, with foliage of a specific dark greenish-brown colour (Fig. 14 and Fig. 15).

Fig. 14: Ventral side of *Batracobdella algira*.
(Present work).

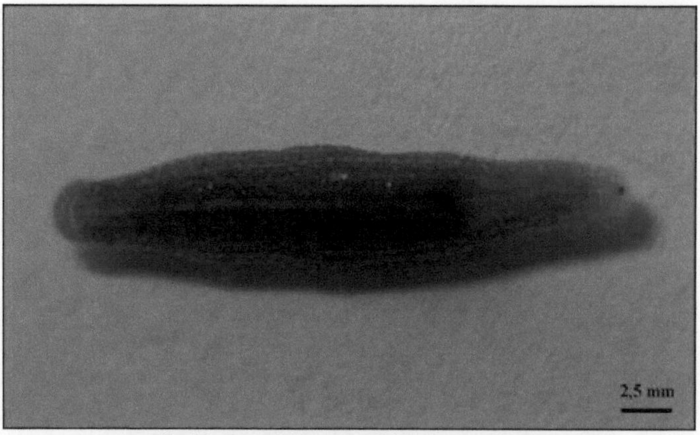

Fig. 15 : Dorsal side of *B. algira* (Present work).

In this species, the eyes are a single cephalic pair and separate from each other (Fig. 16). Histological study revealed the presence of a proboscis characteristic of the order Rhynchobdella. This proboscis has a developed musculature (Fig. 17), which explains its function as a feeder.

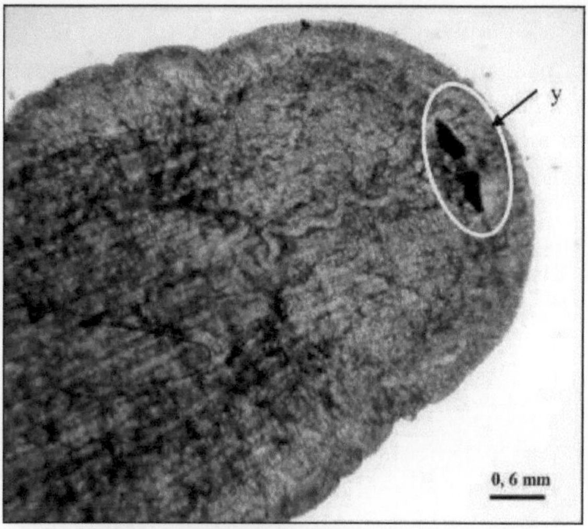

Fig. 16: Eye arrangement in *B. algira.* y: eyes (Present work).

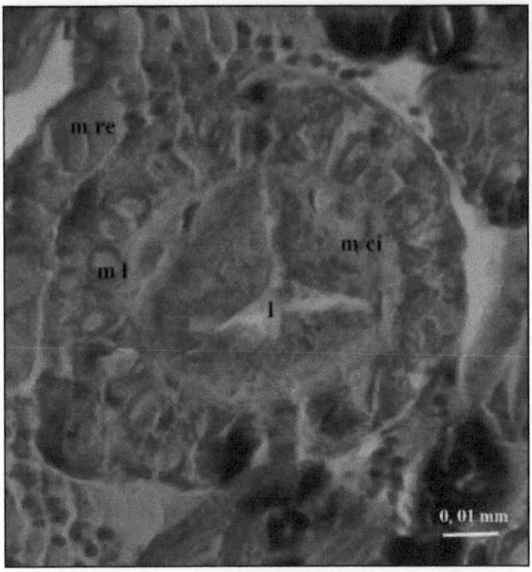

Fig. 17 : Transverse section through the proboscis in *B. algira*.
l: lumen of the tube, m ci: circular muscles, m l: longitudinal muscles, m re: retractor muscles (Present travail).

There are seven pairs of digestive caeca, as seen through the eye in the juvenile stage (Fig. 18). The male and female gonopores are separated by a single ring. Juveniles are attached to their mother's belly (Fig. 19).

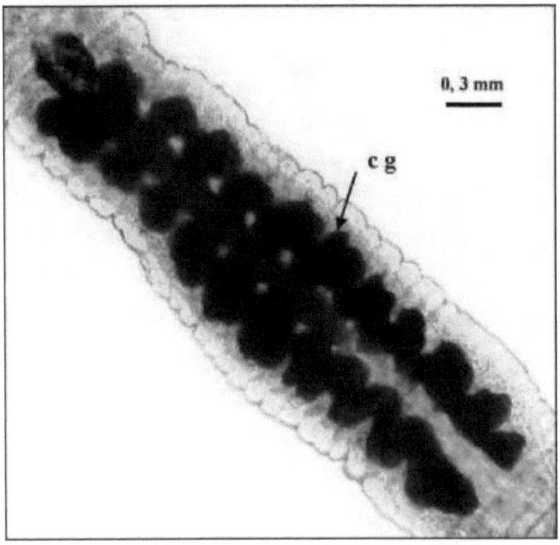

Fig. 18: Gastric caeca in *B. algira*. c g: Gastric caeca (Present work).

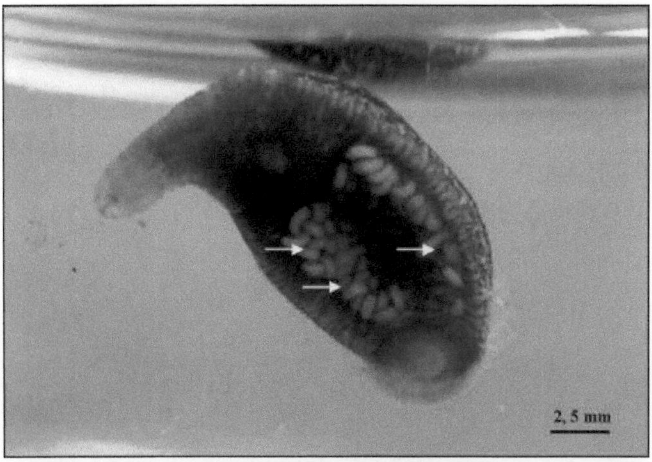

Fig. 19: Ventral side of *B. algira* showing the trophic relationship between the mother and her juveniles (Fleches) (Present work).

In the literature, trophic and protective relationships between mothers and juveniles are mainly noted and studied in vertebrates such as mammals and birds. These relationships are little studied in Hirudineae, and more particularly in the Glossiphoniidae family.

Fig. 20: *Bufo mauritanicus* parasitized by *B. algira* (Present work).

With regard to the feeding regime, according to the literature, *B. algira* feeds by sucking the blood of Anuran amphibians (*Discoglossus pictus, Rana rudibunda...*) and urodeles (*Hydromantes genei*), in the present work we observed it on *Bufo mauritanicus* (Fig. 20). It is therefore a hematophagous species.

1.2. *Theromyzon tessulatum*: (O.F. MUller, 1774)

Hirudo tessulata **O.F. Muller, 1774** (Typical locality: Denmark)

Hirudo tessellata **Bosc, 1802**

Glossiphonia eachana **Thompson, 1846**

Hirudo vitrina **Dalyell, 1853**

Theromyzon pallens **Philippi, 1867**
Hemiclipsis tessellata **Vejdowsky, 1884**
Theromyzon tessulatum **Autrum, 1936**

Theromyzon tessulatum also has a dorsoventrally flattened, almost translucent olive-green foliaceous body with black spots corresponding to the chromatophores (Fig. 21 and Fig. 22).

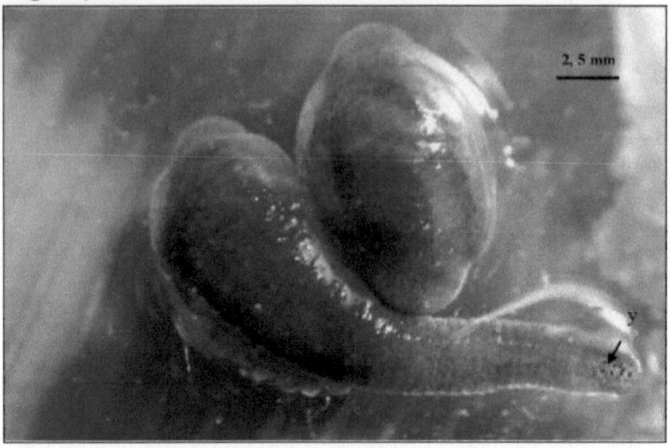

Fig. 21 : Dorsal side of *Theromyzon tessulatum*. y : Eyes (Present work).

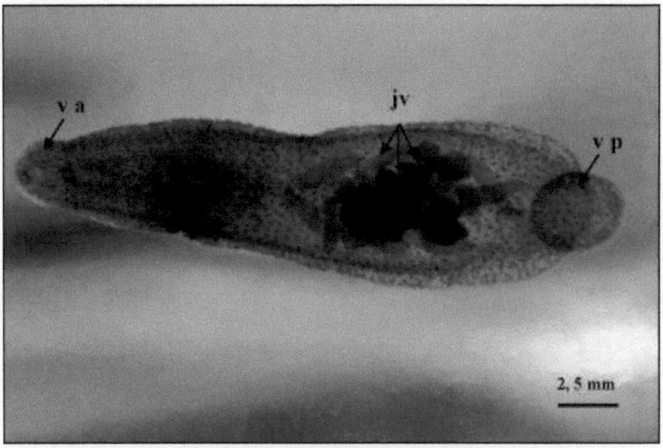

Fig. 22 : Ventral side of *T. tessulatum*.
***v* a: front suction cup, v p: rear suction cup, jv: juveniles (Present work).**

The histological study showed us the presence of a very muscular tube (Fig. 23) and consequently very contractile, playing an important role in feeding.

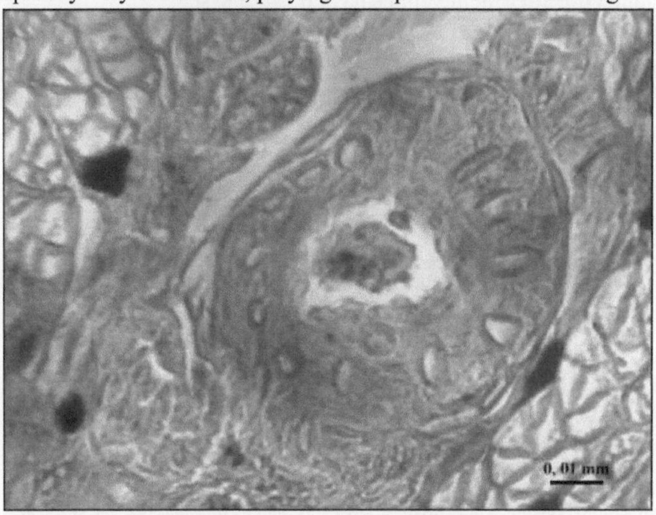

Fig. 23: Histological section of the tube in *T. tessulatum* l: tube lumen, m ci: circular muscles, m l: longitudinal muscles, m re: retractor muscles (Present work).

This species is characterised by four pairs of cephalic eyes arranged in two paramedian lines (Fig. 24).

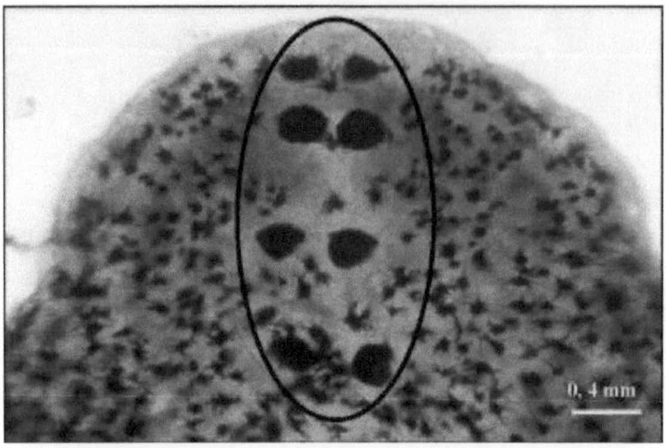

Fig. 24 : Eye arrangement in *T. tessulatum* (Present work).

It should be noted that the eggs and then the juveniles are attached to their mother's belly (Fig. 25) and that this species is considered by **Grasse (1959)** to be prolific, laying 65 to over 600 eggs. This is clearly a reproductive strategy. It reproduces only once in its life and dies (semelparous); this is confirmed in the laboratory, since the adults die after the juveniles have been released.

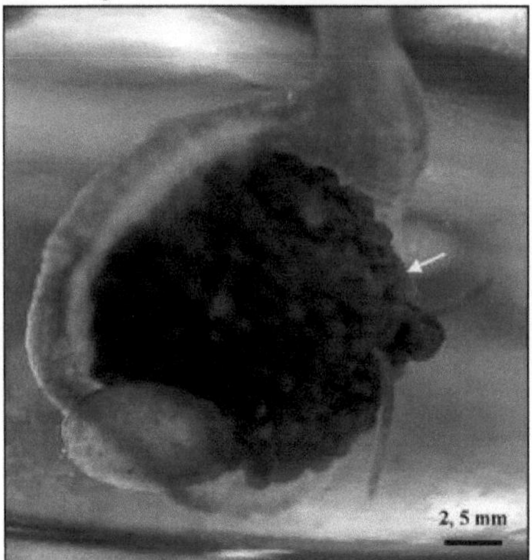

Fig. 25: Trophic relationship between *T. tessulatum* and its juveniles (Fleche) (Present work).

According to **Oosthuizen and Davies (1992),** the identification of species of the genus *Theromyzon* in North America and elsewhere is based on the number of rings separating the gonopores. In *Theromyzo* n *tessulatum*, the gonopores are separated by four rings (Fig. 26).

This species is characterised by the presence of an anticoagulant in its saliva: therostatin **(Chopin *et al.*, 2000 *In* Siddall *et al.*, 2005).**

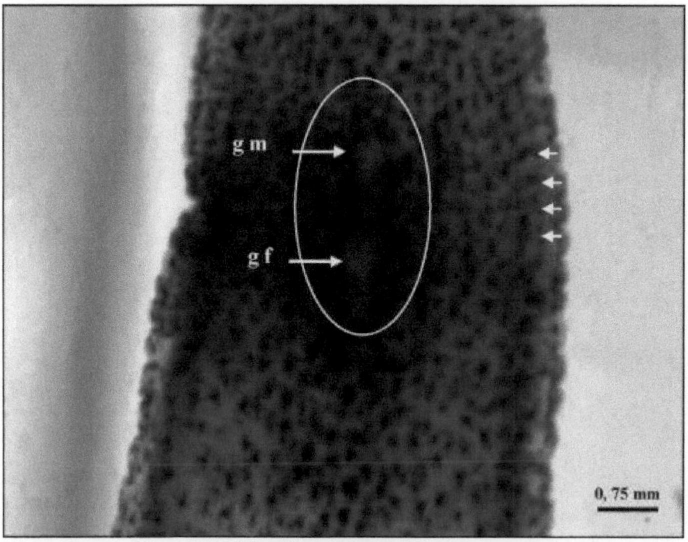

Fig. 26 : Arrangement of the male and female gonopores of *T. tessulatum* on the ventral surface of the animal.
g m: male gonopore, g f: female gonopore (the arrows indicate the rings separating the two gonopores) (Present work).

According to **Grasse (1959),** *Theromyzon tessulatum* changes body colour according to whether it is placed in the dark or in the light, and according to whether it is spread or contracted. According to our observations, this phenomenon is not accompanied by any tendency towards homochromy with the support (Figs. 27, 28 and 29). Indeed, **Smith (1942 *In* Grasse, 1959),** suggests the presence of three types of pigment cells. These occupy positions that vary from one individual to another.

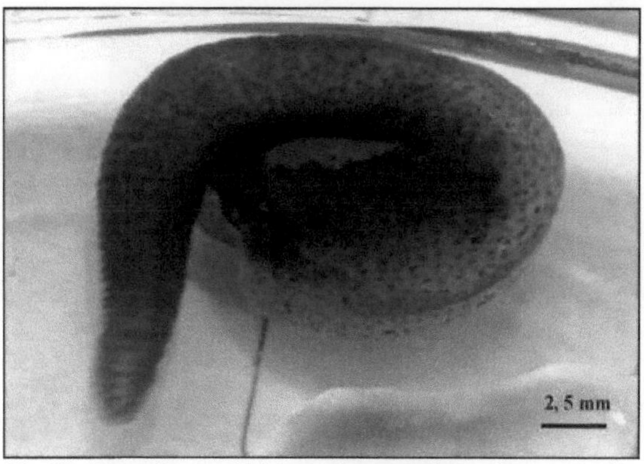

Fig. 27 : *T. tessulatum* with brown chromatophores (Present work).

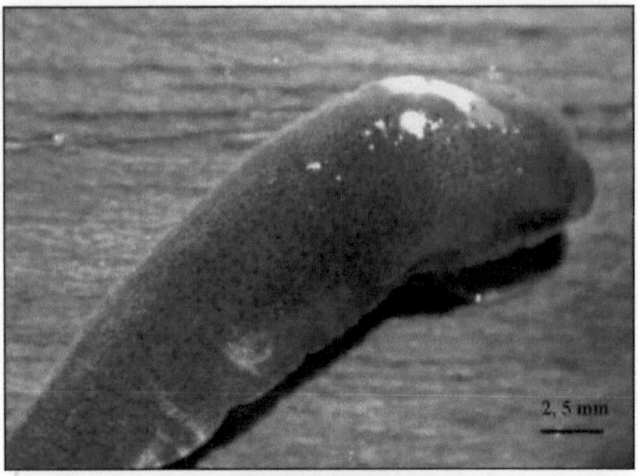

Fig. 28 : *T. tessulatum* with green chromatophores (Present work).

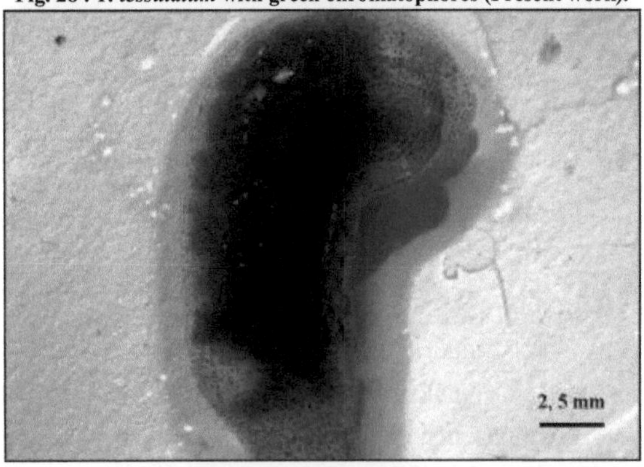

Fig. 29 : *T. tessulatum* with dispersed black chromatophores (Present work).

This species is called "The duck leeche" by the Americans **(Siddall and Budinoff, 2005)**, reflecting a hematophagous diet associated with ducks (Fig. 30).

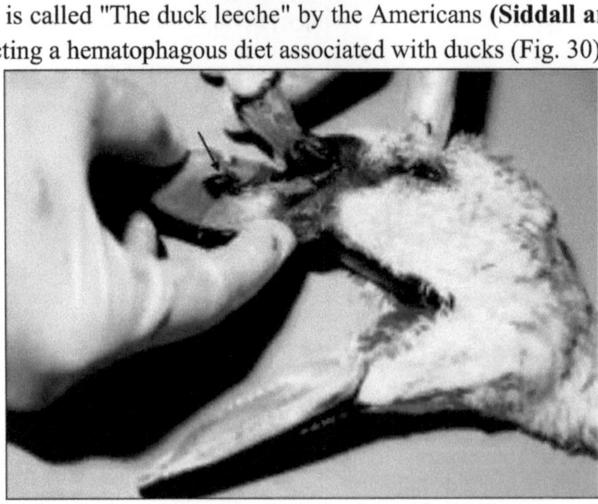

Fig. 30: Nasal cavity of a duck parasitized by *T. tessulatum* (Fleche) (Anonymous 2: website).

1.3. *Placobdella nabeulensis* (Fr. MUller ,1846)

Clepsine costata **Fr. Muller ,1846** (Typical locality: Crimee)

Glossiphonia catenigera **Moquin-Tandon 1846**

Haementeria costata **De Filippi, 1849**

Clepsine affinis **Diesing, 1850**

Placobdella raboti **Blanchard, 1893**

Placobdella guernei **Blanchard, 1894**

Placobdella catenigera **Blanchard, 1894**

Haementeria Nusbaumi **Gedroyc, 1913**

Placobdella costata **Mann, 1961**

This species has a foliaceous body, flattened dorsoventrally, with the organs visible through the glass.

The dorsal side has a greenish-brown background, with three pairs of interrupted black stripes on either side of a broad, black, interrupted median stripe (Fig. 31). This surface is covered with papillae (Fig. 32).

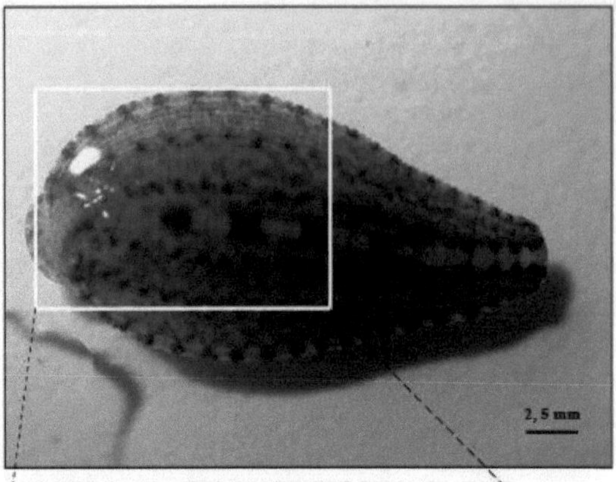

Fig. 31 : Face dorsale de *P. nabeulensis* (Présent travail).

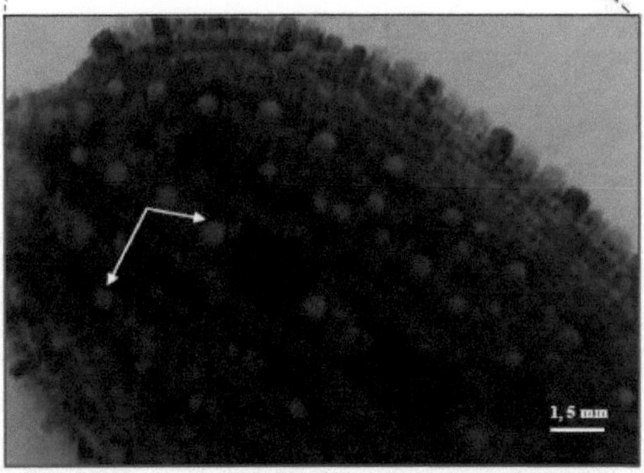

Fig. 32 : Face dorsale de *P. nabeulensis* montrant des papilles (Flèches) (Présent travail).

Fig. 31 : Dorsal side of *P. nabeulensis* **(Present work). **

Fig. 32 : Dorsal surface of *P. nabeulensis* showing papillae (Fleches) (Present work).

In this species, the anterior sucker (Fig. 33) is clearly underdeveloped compared with the posterior sucker (Fig. 34). This is characteristic of the family.

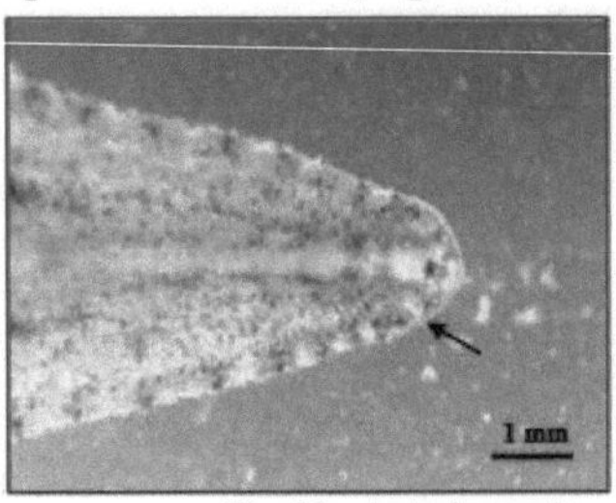

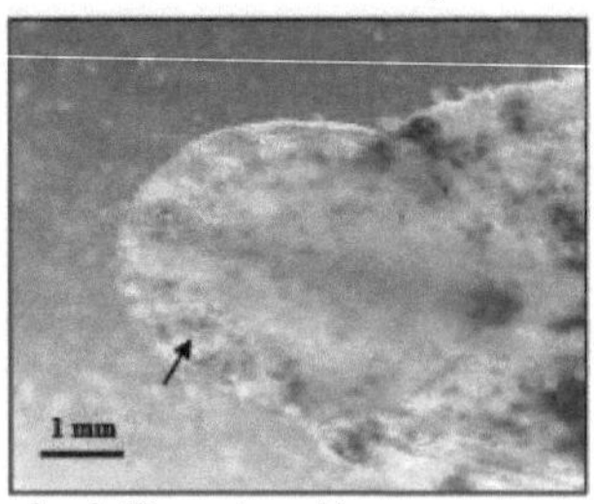

Fig. 33 : Anterior sucker of *P. nabeulensis* (Present work).

Fig. 34 : Posterior sucker of *P. nabeulensis* (Present work).

Unlike *Batracobdella algira*, *Placobdella nabeulensis* has a single pair of non-separated eyes (Fig. 35).

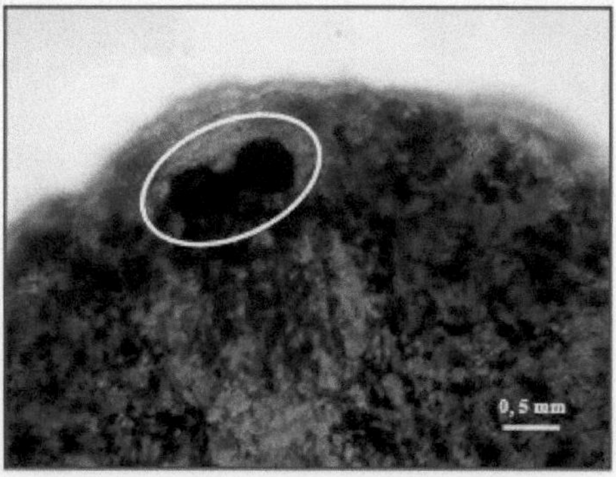

Fig. 35 : Eye arrangement in *P. nabeulensis* (Present work).

Dissection of one specimen (Fig. 36) showed the presence of a proboscis (Fig. 37), two salivary glands and two oesophageal sacs characteristic of the genus *Placobdella*. The latter have recently been termed "mycetomas". According to **Siddall *et al* (2004),** they are considered to be two caecal-like sacs that open into the resophagus halfway between the proboscis and the first pair of gastric caeca. They contain symbiotic bacteria that facilitate the digestion of ingested blood.

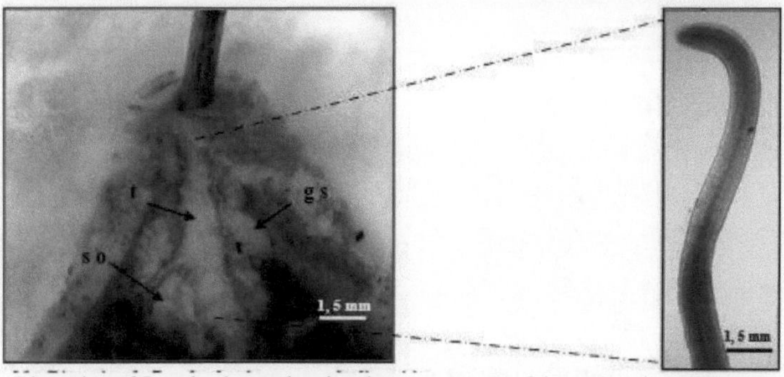

Fig. 36 : Dissection of *P. nabeulensis* showing the arrangement of the proboscis, salivary glands and oesophageal sacs. g s: salivary glands, s o: oesophageal sac, **t: proboscis (Present work).**

Fig. 37 : Trompe of *P. nabeulensis* (Present work).

The gastric caeca in this species consist of seven pairs of lobes, visible through the tube, with six pairs of testis between them (Fig. 38).

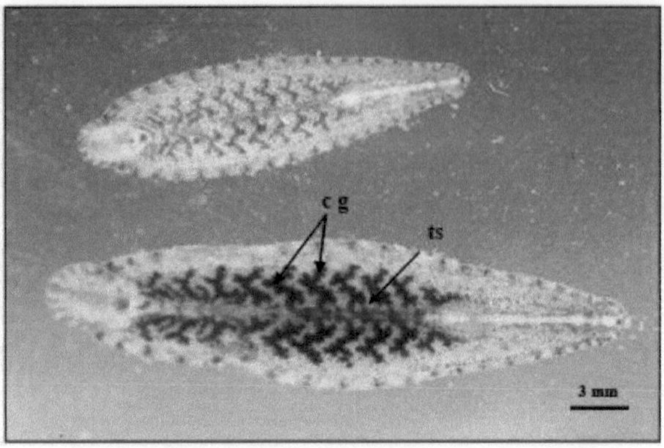

Fig. 38 : Ventral side of *P. nabeulensis.* cg : gastric caeca, ts : testis (Present work).

The male and female gonopores of *Placobdella nabeulensis* are separated by two rings and open in the clitellar region (Fig. 39). The male gonopore always precedes the female. There is a single pair of ovaries located in front of six pairs of testes.

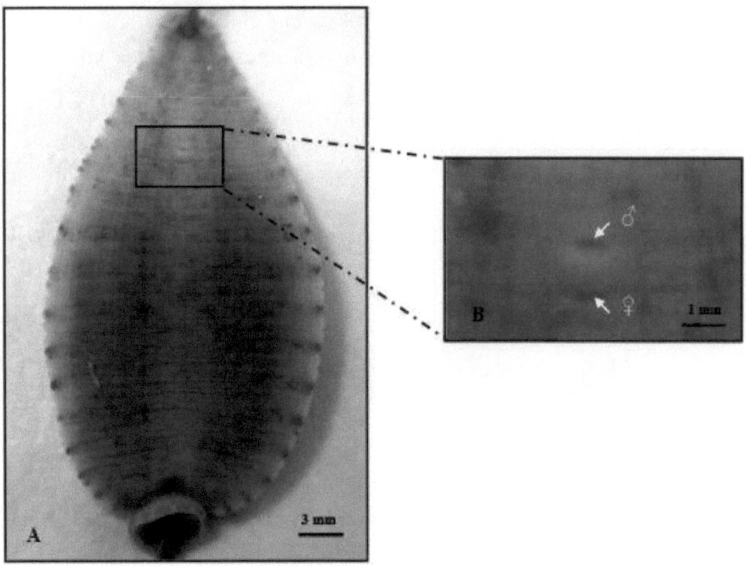

Fig. 39 A and B: Gonopore arrangement on the ventral side of *P. nabeulensis* (Present work).

This species also shows a physiological change in body colour (Fig. 40). Indeed, **Sawyer and Davies (1974)** have shown that light has a major effect on the colour change of certain leeches. They turn brown in the light and lighten in the dark.

Fig. 40 : *P. nabeulensis* in different colours (Present work).

Dissection of a specimen enabled us to observe the layout of the gastric caeca and the nerve chain (Fig. 41).

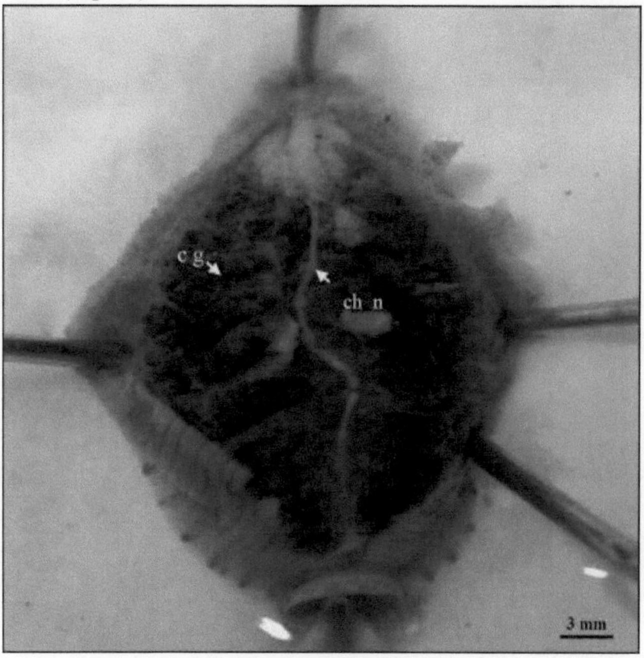

Fig. 41: Dissection of *P. Nabeulensis* showing the nerve chain and gastric caeca. ch n: nerve chain, c g: gastric caeca (Present work).

Species of the genus *Placobdella* attack humans, mammals and reptiles, particularly turtles. *Placobdella costata* was collected by Minelli (1979) from *Emys orbicularis*. In the present work, *P. nabeuelensis* was observed on the freshwater turtle *Mauremys leprosa* (Figs. 42 and 43).

Fig. 42 : *Mauremys leprosa* (Present work).

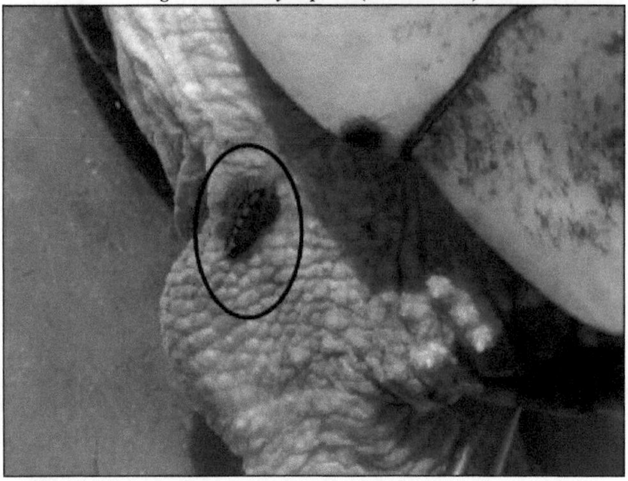

Fig. 43: *Mauremys leprosa* parasitized by *P. nabeulensis* (Present work).

1.4.*Alboglossiphonia hyalina* (Linnaeus, 1758)

Hirudo heteroclita **Linnaeus, 1761** (Lacking a precise typical locality it was first described in Europe by Linnaeus).

Hirudo hyalina **O.F. Muller, 1774**

Hirudo swampina **Bosc, 1802**

Hirudo papillosa **Braun, 1805**

Hirudo trioculata **Carena, 1820**

Hirudo arcuata **Fabricius, 1826**

Clepsine hyalina **Moquin-Tandon, 1827**

Clepsine carenae **Moquin-Tandon, 1827**

Clepsina hyalina **Brightwell, 1842**

Glossiphonia heteroclita **Moquin-Tandon, 1846**

Glossopora hyalina **Johnston, 1846**

Glossiphonia carenae **Moquin-Tandon, 1846**
Glossiphonia arcuata **Moquin-Tandon, 1846**
Clepsine swampina **Diesing, 1850**
Clepsine papillosa **Grube, 1851**
Glossiphonia hyalina **Thompson, 1856**
Clepsine heteroclita **Whitman, 1878**
Clepsine striata **Apathy, 1888**
Glossiphonia trioculata **Blanchard, 1893**
Glossiphonia (Alboglossiphonia) heteroclita **Harding, 1910**

Alboglossiphonia hyalina has a foliaceous body, flattened dorsoventrally, with its organs visible through the glass. It is light yellow in colour (Fig. 44). There are no papillae on the dorsal surface of this species.

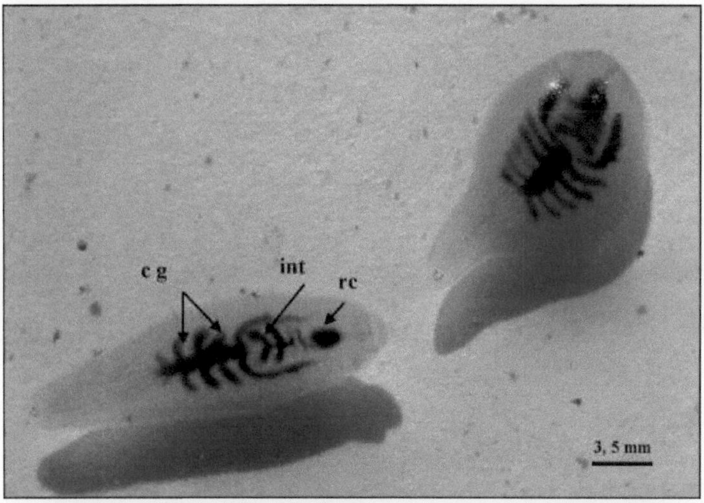

Fig. 44: Dorsal surface of *Alboglossiphonia hyalina* c g: gastric caeca, int: intestine, rc: rectum (Present work).

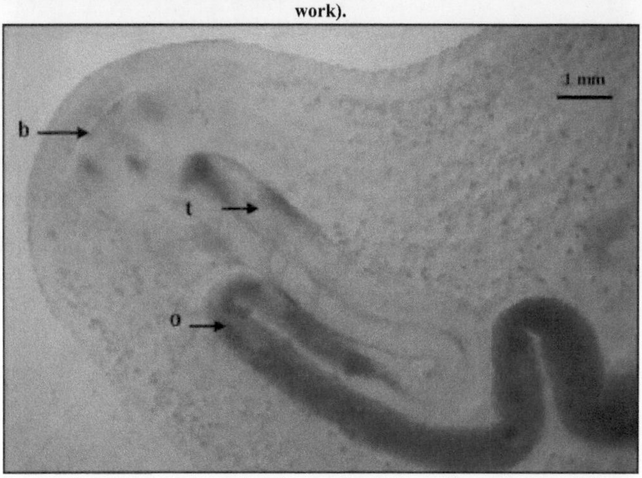

Fig. 45 : Ventral side of *A. hyalina* showing the junction between the proboscis and the resophagus.

b: mouth , t: tube , o: oesophagus (Present work).

The mouth (Fig. 45) is in the middle of the front sucker and not on its edge. The proboscis of this species is clearly visible through the tube. Even the gastric caeca (Fig. 44), of which there are six pairs, are simple and not lobed. There are three pairs of eyes (Fig. 46), grouped together in the shape of a triangle.

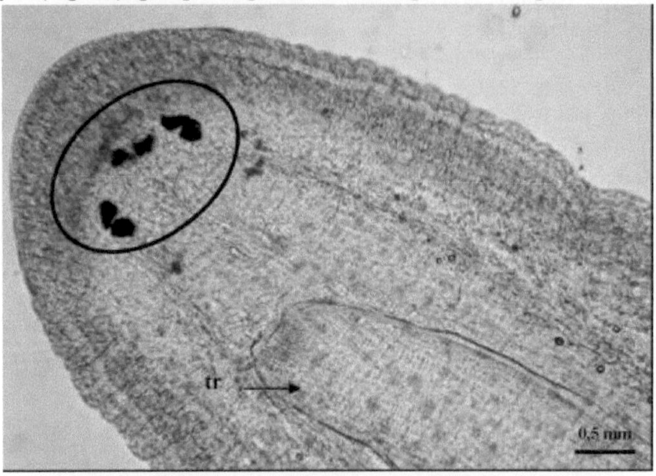

Fig. 46 : Eye arrangement in *A. hyalina.* tr : proboscis (Present work).

The male and female genitalia have a common gonopore (Fig. 47).

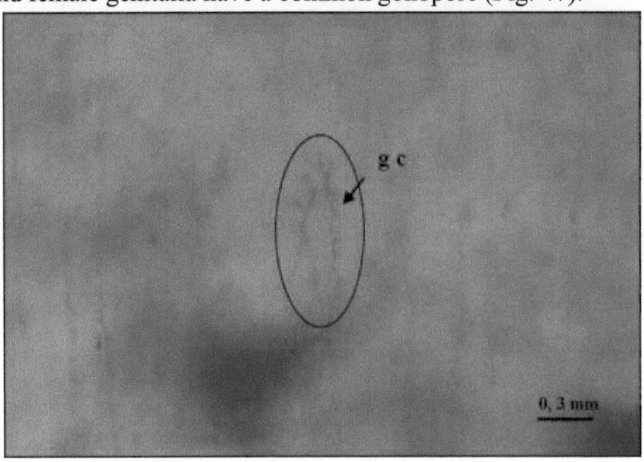

Fig. 47 : Common gonopore in *A. hyalina* g c : Common gonopore (Present work).

The juveniles are also attached to their mother's belly (Fig. 48), ensuring the secretion of yolk, which they will use as food **(Grasse, 1959)**.

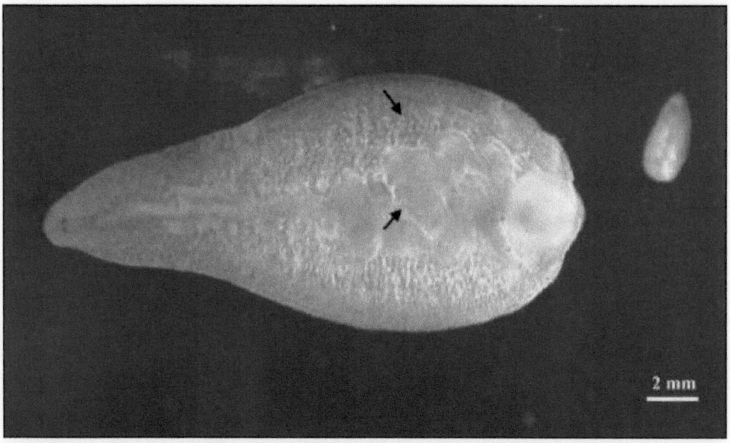

Fig. 48: Trophic relationship between *A. hyalina* and its juveniles (Fleches) (Present work).

According to the literature, *A. hyalina*, still called "the snail leeches" by the Americans, is a liquidostomatophagous species (sucking mollusc fluids). It feeds mainly on Gasteropods **(Govedich *et al.*, 2004).**

2. Family Hirudinidae Whitman, 1886

The main characteristic of this family is that the mouth has three serrated jaws (Trignathes) that can cut the skin or mucous membranes. As a result, these leeches lead an excellent parasitic life. They are exclusively hematophagous. The absence of the proboscis makes it possible to place this family within the order Arhynchobdella.

The body is vermiform, of varying length, and the spongy cocoons, of varying size, are left on the substrate without any parental protection.

The only genus of this family found in the Lebna dam is the *Hirudo* genus, of which four species have been collected. These differ in body coloration which, according to **Hetchel and Sawyer (2002), is** the most useful and reliable criterion for distinguishing medicinal species of this genus. Internal morphology, and more specifically comparison of the reproductive systems, is only used if the external colouration has been damaged by a fixative, for example. We have identified the species *Hirudo troctina* and named the other three respectively: *Hirudo sp.1*, *Hirudo sp.2* and *Hirudo sp.3*.

These last three are so named because they are distinguished by criteria that do not have their counterparts in the other determination keys. Further phylogenetic studies will enable us to decide whether these are new species or not.

2.1. Genus *Hirudo* Linne, 1785

All the captured species of the family Hirudinidae have the characteristics of the genus *Hirudo*. There are three jaws (Trignathes) (Fig. 49 A). They are armed with a single row of sharp teeth (monostichodonts) (Fig. 49 B and C), the number of which varies from 60 to 100. The ventral surface of the upper yeast does not have a longitudinal fissure (Fig. 50).

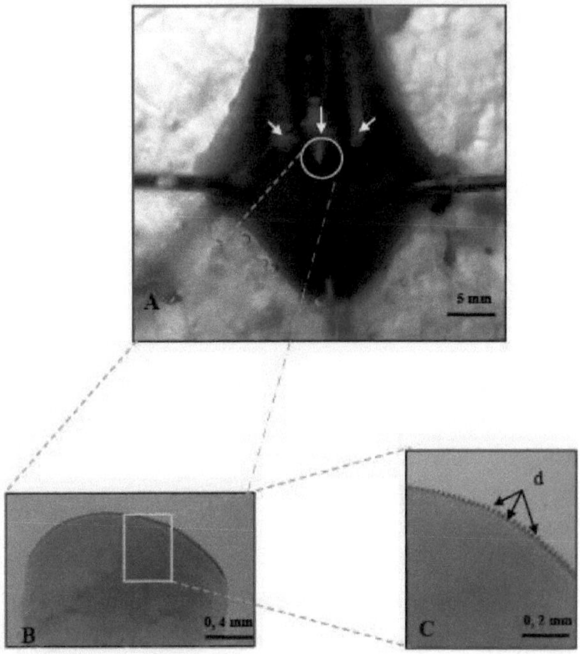

Fig. 49 : Jaws of *Hirudo troctina.*
***A*: trignathic jaws (Fleches); B and C: details of a jaw showing a single row of teeth (monostichodonts); d: teeth (Present work).**

In addition to these two main characteristics of the genus *Hirudo*, *these* species also share other common criteria found in other genera (*Haemopis* and *Limnatis*), such as the number of eyes and the number of rings separating the male and female gonopores.

These last two genera (*Hemopis and Limnatis*) are distinguished from the genus *Hirudo* by the presence of a labial furrow in *Limnatis*, whereas two rows of teeth (Distichodontes), which are few in number, are present in species of the genus *Haemopis.* It should be noted that these two genera are reported in the literature for Tunisia **(Moore, 1939).**

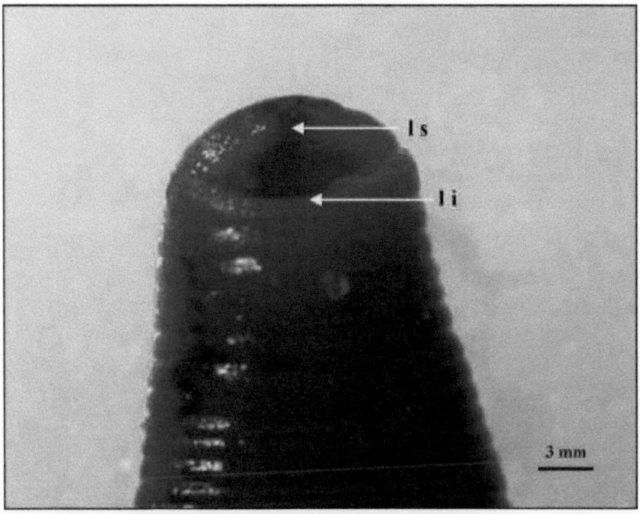

Fig. 50 : Anterior sucker in *Hirudo troctina* showing the absence of a labial fissure at the level of the upper lip. l s : upper lip, l i : lower lip (Present work).

In the genus *Hirudo*, there are five pairs of eyes arranged in a parabolic arc on the first segments of the anterior sucker (Fig. 51). In this case, the eyes are marginal, unlike in the Glossiphoniidae family, which are cephalic.

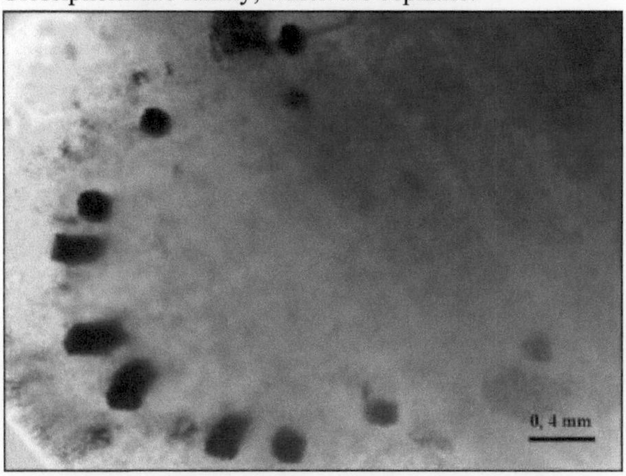

Fig. 51 : Marginal arrangement of the eyes in *Hirudo troctina* (Present work).

The male and female gonopores are separated by five rings (Fig. 52). Note that the male gonopore is more visible and always precedes the female.

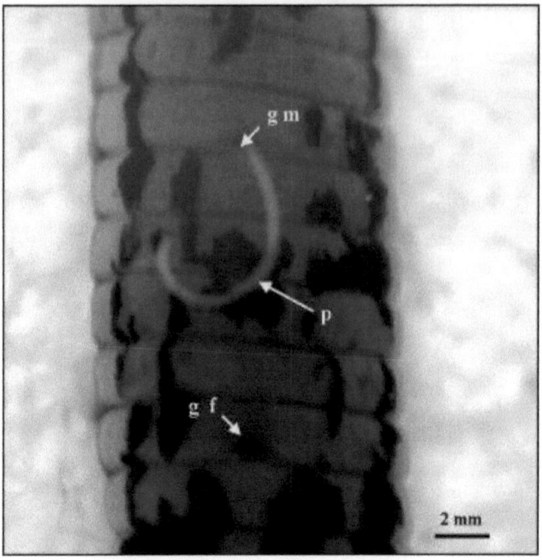

Fig. 52: Position of male and female gonopores in *Hirudo troctina* g m: male gonopore, g f: female gonopore, p: penis (Present work).

2.1.1 *Hirudo troctina* (johnson, 1816)

Hirudo troctina **Johnson, 1816** (no specific typical locality)
Sanguisuga interrupta **Moquin-Tandon, 1827**
Sanguisuga troctina **Moquin-Tandon, 1846.**

This species has a vermiform cylindrical body. The dorsal surface is greenish, with two thin yellow-orange paramedian stripes intersected by quadrangular spots of the same colour. There are also two thin paramarginal bands of the same colour as the previous ones. These are intersected by black quadrangular patches and two wide, yellow-orange lateral bands intersected by black quadrangular patches (Fig. 53).

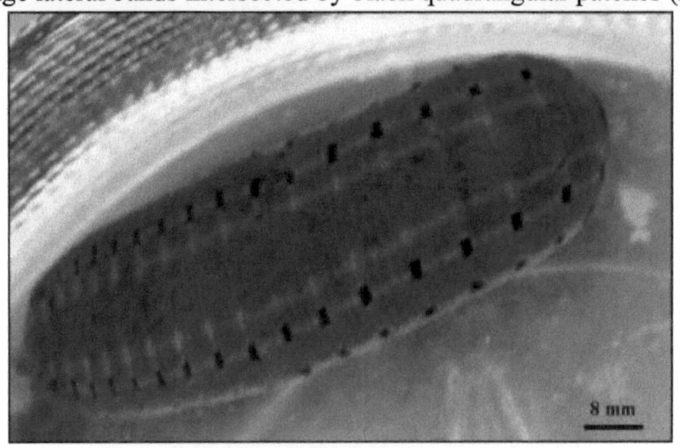

Fig. 53 : Dorsal side of *Hirudo troctina* (Present work).

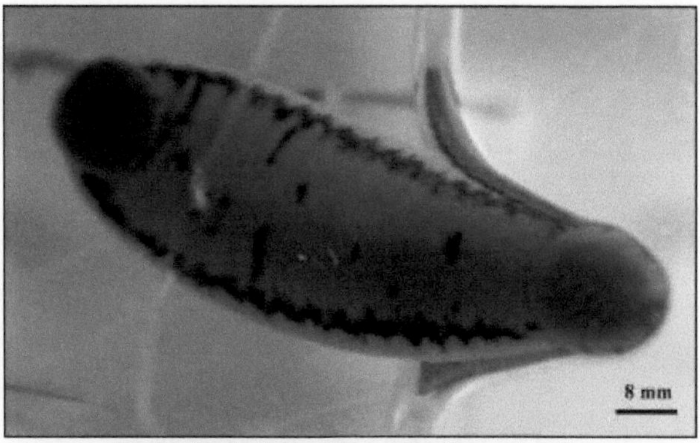

Fig. 54 : Ventral side of *H. troctina* (Present work).

The ventral side (Fig. 54) is light green with large black spots. There is also a black zigzag outline. It should be noted that this last criterion distinguishes this African species from the Australian species *Hirudo orientalis* **(Utevsky et Trontelj, 2005).** *Hirudo troctina*, like the whole family, is characterised by the absence of any parental protection of juveniles. In fact, dozens of more or less voluminous spongy cocoons are abandoned in a humid environment very close to the water (Fig. 55, A and B). This observation was demonstrated in the laboratory. *H. troctina* in elevage has chosen to place its cocoon fixed to the wall of the elevage bottle and not immersed in water.

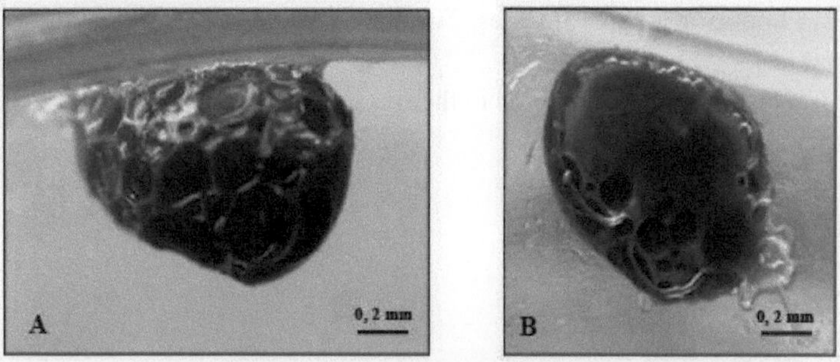

Fig. 55: Cocoons of *Hirudo troctina*, A: dorsal side, B: ventral side (Present work).

According to **Baker and Macagno (2001),** the cocoon diameter of *Hirudo medicinalis* varies between 10 and 15 mm. In the present work we found that the cocoon diameter of *H. troctina* is smaller and varies between 0.8 and 10 mm.

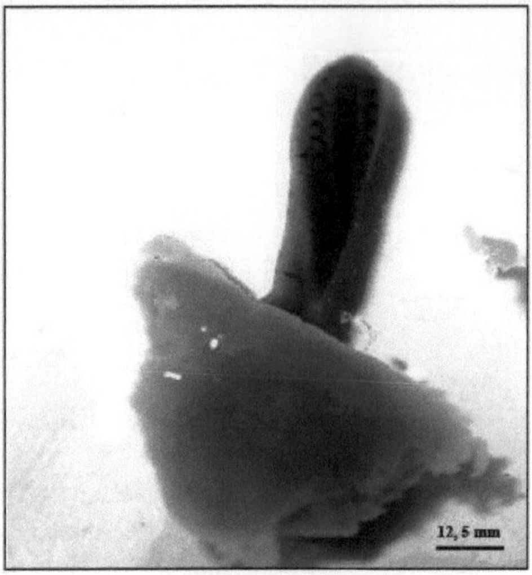

Fig. 56 : *H. troctina* sucks blood from a piece of liver (Present work).

H. troctina is a hematophagous species, feeding on the blood of various mammals (cows and horses), amphibians and reptiles (Fig. 56).
Dissection of a specimen *of Hirudo troctina* has enabled us to see the morphology of the reproductive system (Fig. 57). This comprises a single pair of ovaries situated in front of ten pairs of testes. The male and female genital pores are single and separate.
The female organs are concentrated in the space between the male orifice and the first pair of testicles.

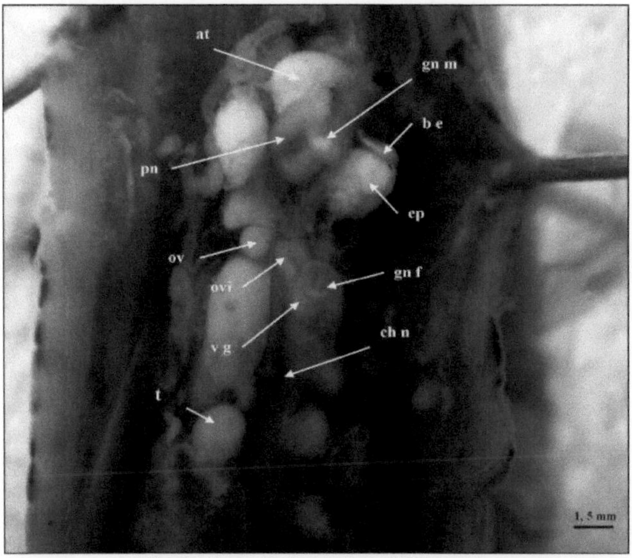

Fig. 57: Reproductive system of *H. troctina*.
at: atrium, b e: ejaculatory bulb, ep: epididymis, gn m: male gonopore, gn f: female gonopore, ov: ovary, ovi: oviduct, pn: penis, vg: vagina, t: testis, ch n: nerve chain (Present work).

Hirudo troctina is characterised by a massive epididymis (Fig. 58), also known as the seminiferous vesicle. The epididymis has a curled region in relation to the ejaculatory bulb, which has a swollen, muscular region. The vagina is bulbous, straight and not wrinkled. The atrium continues into a long penis.

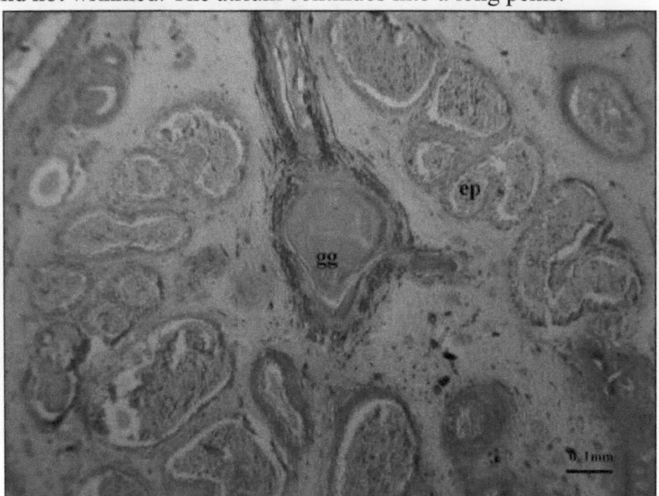

Fig. 58: Sagittal section showing the epididymis with a pelotonous structure in *H troctina*. ep: epididymis, gg: nerve ganglion (Present travail).

2. 1.2. *Limnatis. sp.1* :

Like all Hirudinidae, this species has a vermiform cylindrical body. The dorsal surface of the animal (Fig. 59) is dark brown with three broad orange stripes running the length of the animal: one in the middle and two on the sides. The ventral side is of a uniform dark brown colour (Fig. 60).

Fig. 59: Dorsal surface of *Limnatis sp.1* (Present work).

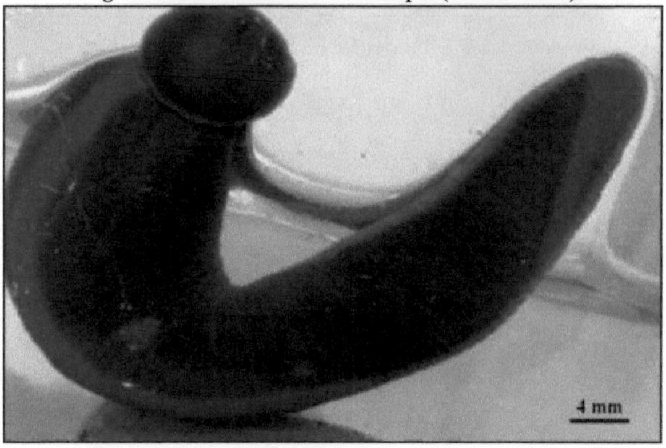

Fig. 60: Ventral side of *Limnatis sp. 1* (Present work).

The dissection of a specimen enabled us to see the reproductive system of this species.

Comparison of the reproductive system of *Hirudo sp.1* with that of *Hirudo troctina* shows that they differ, especially in the shape of the epididymides. *Hirudo sp.1*'s epididymides are more massive and more rounded.

The difference in external coloration, not considered a fundamental criterion for specific determination in other zoological groups, is in this case the most reliable criterion. This difference in colour is confirmed by the difference in the organisation of the reproductive system (Fig. 61).

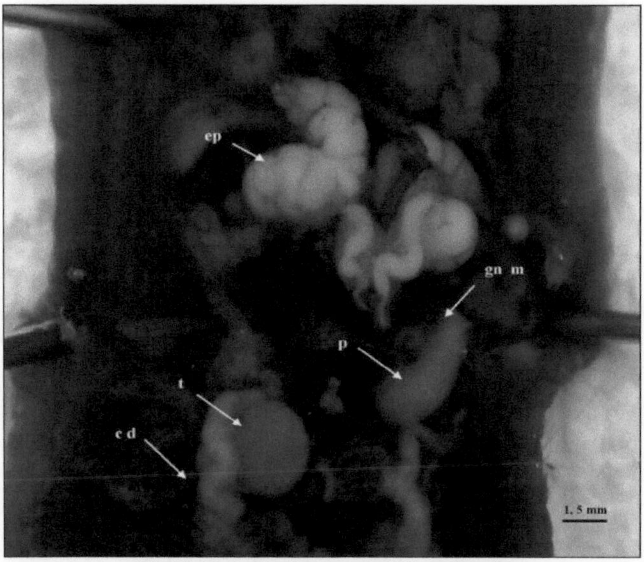

Fig. 61 : Reproductive system of *Limnatis sp.1*.
ep: epididymis, gn m: male gonopores, p: penis, t: testis, c d: deferent canal (Present work).

2. 1.3. *Limnatis. sp.2*

Hirudo sp.2 has a vermiform cylindrical body, the dorsal surface of which is a non-uniform dark green colour with three orange stripes, one median and two lateral. There are also four light lines on either side of the median band (Fig. 62). The ventral side is uniformly brown (Fig. 63).

Fig. 62 : Dorsal side of *Limnatis. sp.2* (Present work).

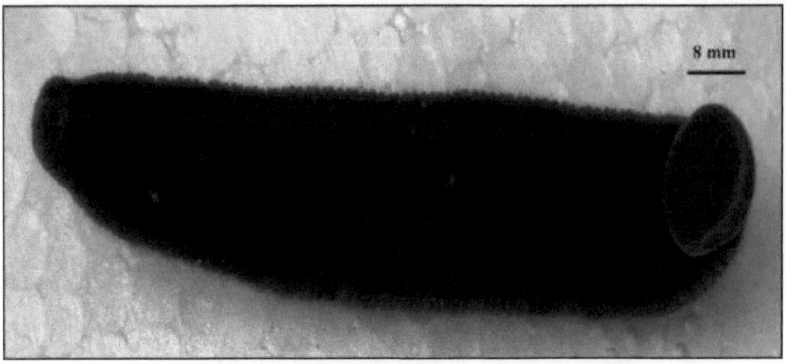

Fig. 63: Ventral side of *Limnatis. sp.2* (Present work).

Dissection of a specimen of *Hirudo sp.2* confirms the difference in the reproductive system, especially the shape of the epididymis, in comparison with *Hirudo troctina* (Fig. 64).

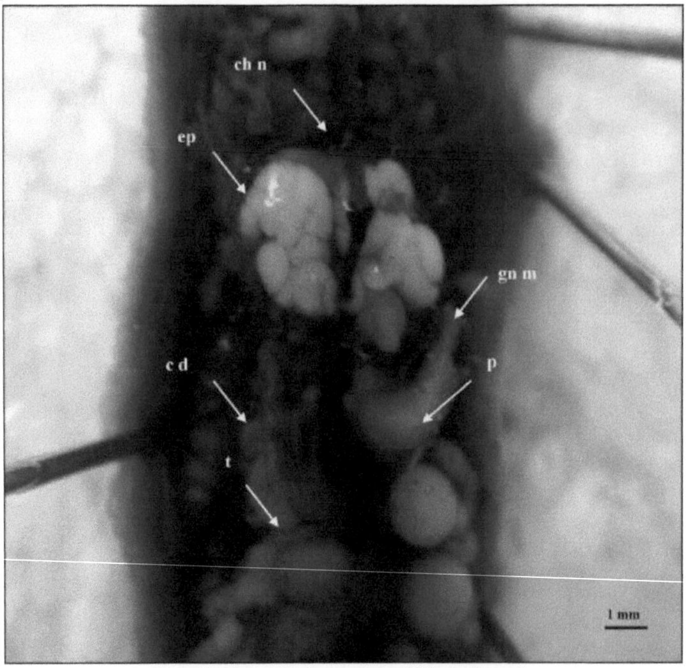

Fig. 64 : Reproductive system of *Limnatis. sp.2*.
ch n: nerve chain, ep: epididymis, gn m: male gonopore, c d: deferent canal, p: penis, t: testis (Present work).

2. 1.4. *Limnatis nilotica*

Hirudo sp.3 has a vermiform cylindrical body with a greenish-brown dorsal surface (Fig. 65) and two orange lateral stripes (Fig. 66). There are 4 longitudinal blackish dotted lines on the dorsal surface. The ventral side is of a uniform brown colour. This species resists well when young; in fact it has been able to survive in the laboratory for ten months without feeding. Dissection of one specimen shows the absence of testis, which could be explained by the fact that this animal uses certain internal organs and tissues for food when subjected to the wrong conditions.

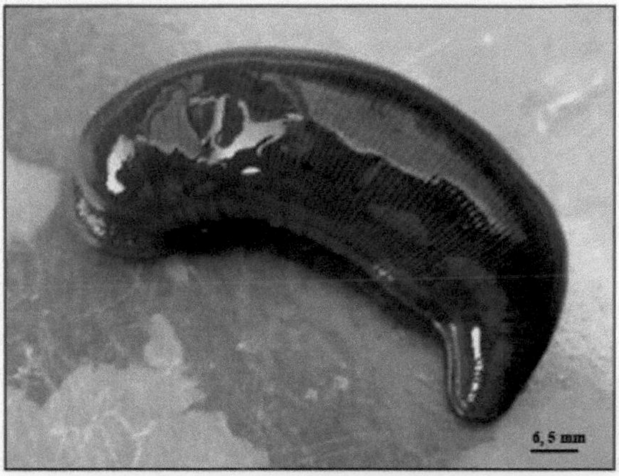

Fig. 65 : Dorsal side of *Limnatis nilotica* (Present work).

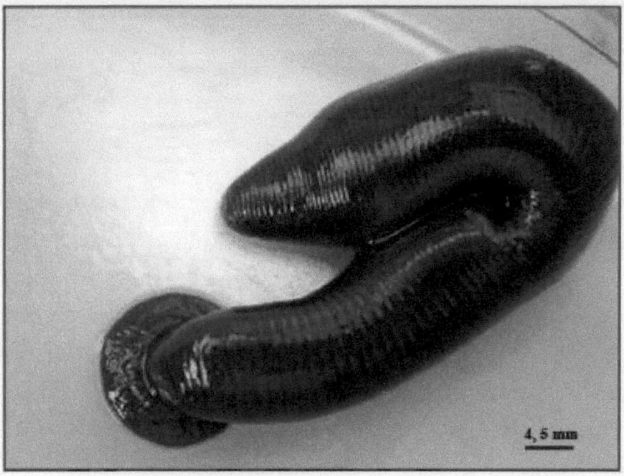

Fig. 66 : Lateral view showing the orange stripe in *Limnatis nilotica* (Present work).

Hirudo sp.3 also shows differences in the reproductive system (Fig. 67) compared with the other species of the genus *Hirudo*. The epididymis is larger than that of *Hirudo sp.1* and *Hirudo sp.2*.

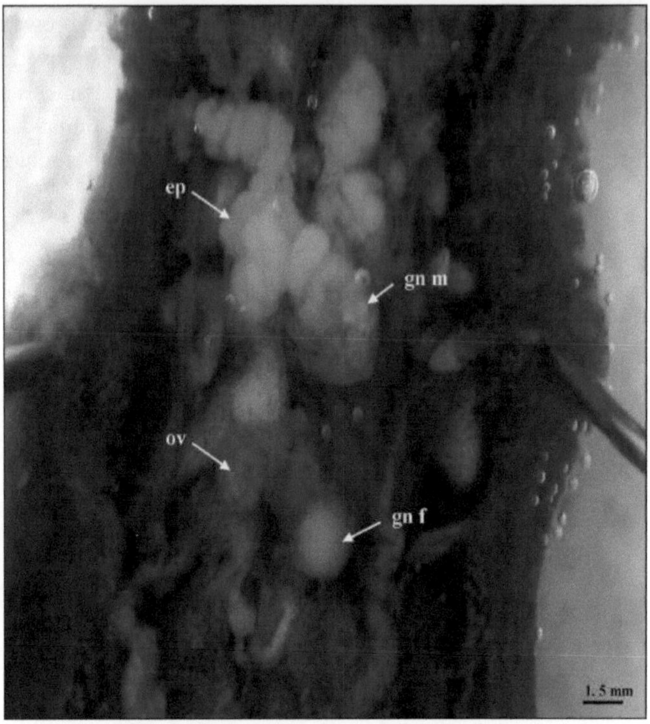

Fig. 67 : Reproductive system of *Limnatis nilotica*
ep: epididymis, gn m: male gonopore, gn f: female gonopore, ov: ovary (Present work).

Note: Taxonomic position

Despite the high conservation value, medical attention and biological interest of medicinal leeches, there has been confusion over their taxonomic position. Historically, medicinal leeches have been considered to be different forms or varieties of the same species *Hirudo medicinalis*, with the exception of *Hirudo troctina*, which was considered a separate species from the outset. The difference in external coloration is attributed to the distinct geographical origins of these forms and has only been valid for *Hirudo troctina* from North Africa, still called by **Brumpt (1949**) "Dragon d'Alger". The distinction between *Hirudo troctina* and *Hirudo medicinalis* was confirmed by **Hetchet and Sawyer (2002)** by comparing the two reproductive systems. They also emphasised external coloration as a reliable distinguishing criterion. **Kutschera (2004)** noted the absence of cross-fertilisation between *Hirudo verbana* and *Hirudo medicinalis* and concluded that they were two different species. A subsequent molecular study by **Trontelj and Utevsky (2004)** was able to settle and confirm these opinions.

In summary, five species belonging to the genus Hirudo are reported in the bibliography: *Hirudo medicinalis* known from western and central Europe, Ukraine and Luthuania; *Hirudo verbana* : Eastern Mediterranean region, Balkans, Moldova, Ukraine, Russia and Armenia; *Hirudo troctina*: Morocco, Tunisia and Algeria; *Hirudo orientalis*: Georgia, Armenia, Azerbaijan and Iran (**Grosser and Pesic, 2006)**; while *Hirudo nioponia* is known from Japan.

3. Family Erpobdellidae Blanchard, 1894

The species in this family are carnivores, predators of aquatic insects. They have abandoned the hematophagous diet of their ancestors **(Apakupakul *et al.*, 1998)**. They are characterised by the absence of parental protection for juveniles, and hundreds of smooth, brown cocoons are subsequently abandoned on various substrates such as pebbles.

Species in the Erpobdellidae family are considered to be good indicators of freshwater pollution **(Wicklum and Davies, 1996 *In* Siddall 2002).** These species are abundant in polluted environments rich in organic substances.

In the present work, we have found only one species of this family in the Lebna dam: *Erpobdella testacea* :

3.1. *Erpobdella testacea* : (Savigny, 1822)

Nephelis testacea **Savigny, 1820** (typical l.ocalite: Paris)

Nephelis octoculata **Blanchard, 1893**

Herpobdella testacea **Johansson, 1910**

Erpobdella testacea **Bennicke, 1943**

Like the species of the Erpobdellidae family, *Erpobdella testacea* has a cylindrical body. Its body colour is a uniform dark brown (Fig. 68).

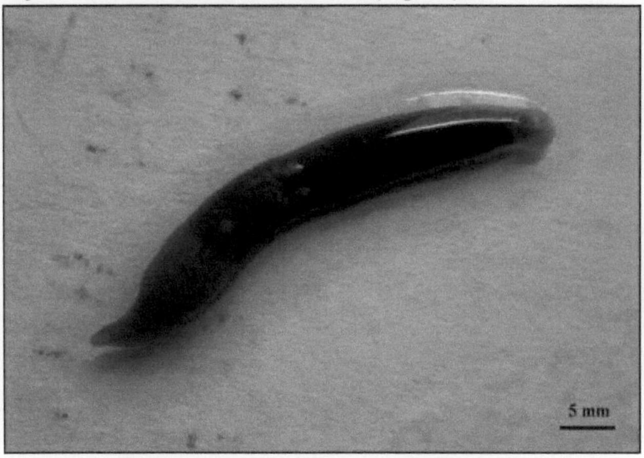

Fig. 68 : Ventral side of *E. testacea* (Present work).

This Hirudinea has a wide mouth that occupies the whole of the anterior sucker. The jaws are absent (Fig. 69).

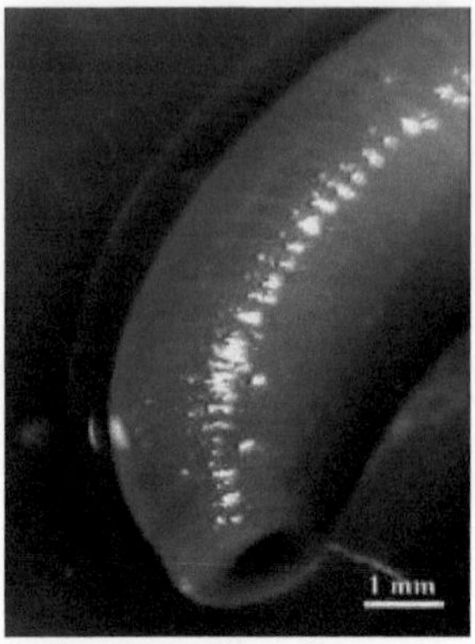

Fig. 69 : Anterior sucker of *E. testacea* (Present work).

There are eight eyes, divided into three groups: a frontal group with four eyes and two lateral groups, each with two eyes (Fig. 70).

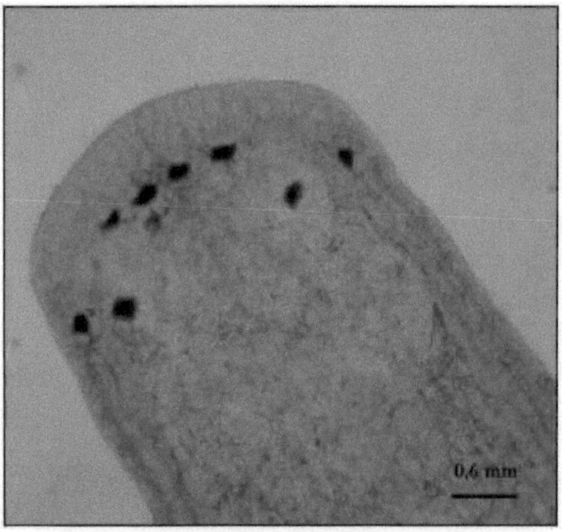

Fig. 70 : Eye arrangement in *E. testacea* (Present work).

The male and female gonopores are separated by four rings (Fig. 71) and the digestive caeca are absent (Fig. 72).

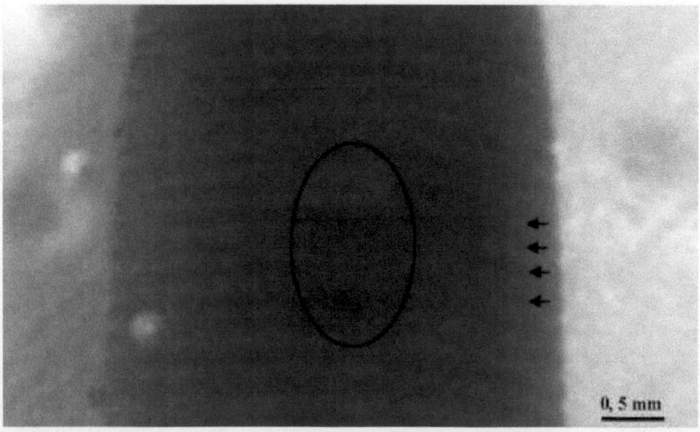

Fig. 71: Position of the male and female gonopores in *E. testacea.* Each arrow corresponds to a ring (Present work).

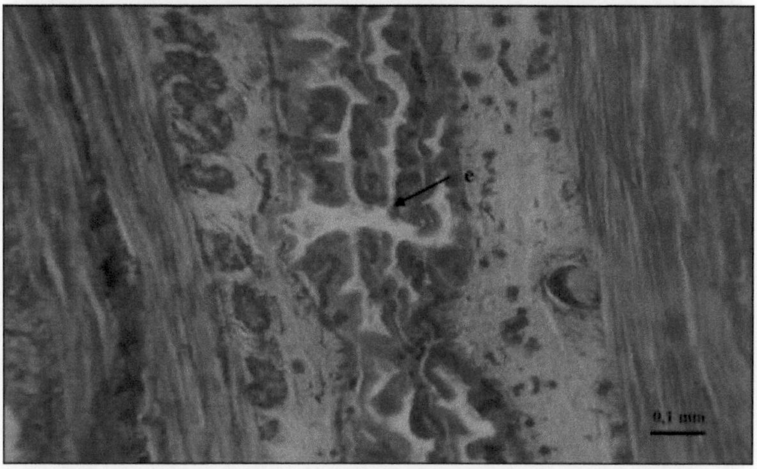

Fig. 72: Sagittal section through the gastric region in *E. testacea* showing the absence of caeca; e: stomach (Present work).

In this family, the proboscis and jaws are absent, making parasitism impossible. Erpobdellidae swallow their prey whole, using a muscular pharynx without jaws **(Govedich *et al.*, 2002).**

At the Laboratory, juveniles were photographed devouring a piece of liver (Fig. 73).

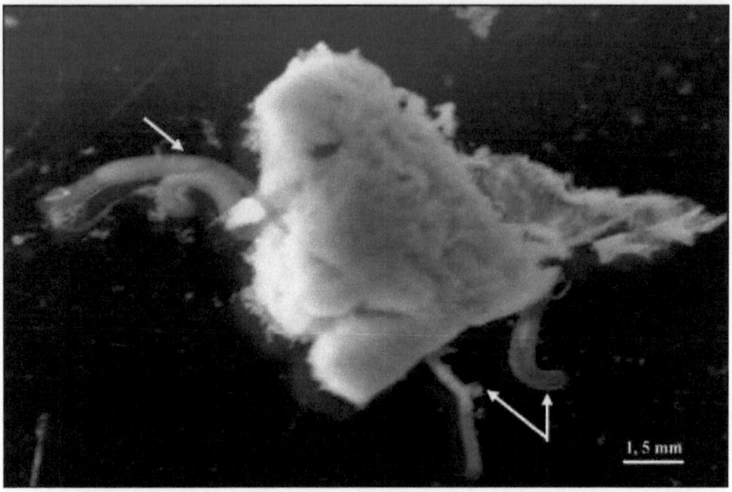

Fig. 73: Juveniles of *E. testacea* feeding on a piece of liver (Fleches) (Present work).

From our observations in the field, we were able to verify the absence of parental protection of juveniles in this species. Hundreds of abandoned cocoons were observed on the substrate (Fig. 74).

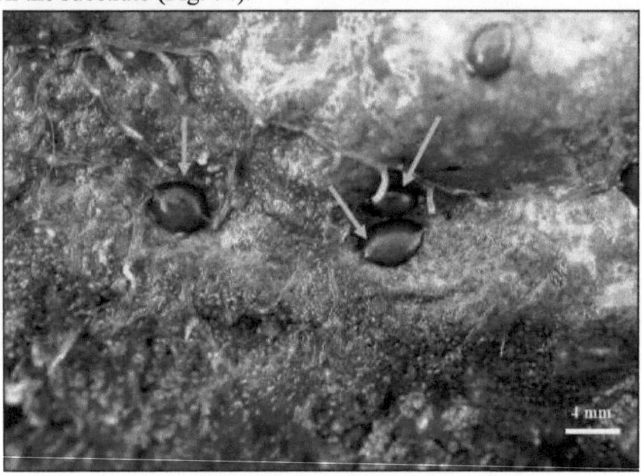

Fig. 74 : Cocoons of *E. testacea* (Fleches) (Present work).

The scale-like cocoon of *Erpobdella testacea* is oval, smooth, flattened and brown in colour. It is almost opaque and filled with a viscous solution that serves as food for the small eggs inside (Fig. 75).

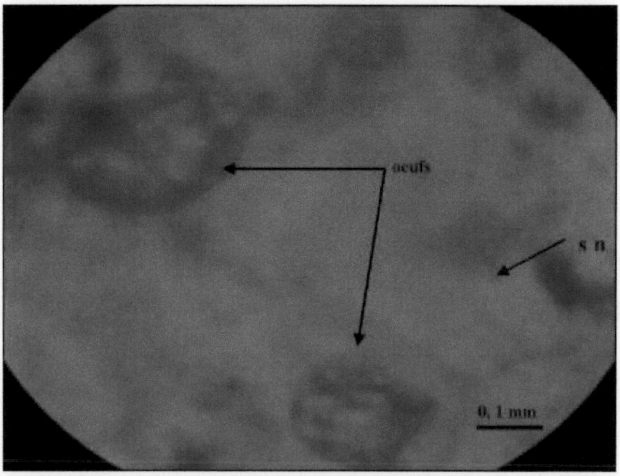

Fig. 75: Cocoon of *E. testacea.* s n: nutrient solution (Present work).

4. Proposal for a general classification of the Hirudinees of the Lebna dam

In Table 2 we propose a classification of Hirudinees based on morphological criteria, feeding regime and the absence or presence of parental protection of cocoons and juveniles.

Table 2 (Present work).

CLASSIFICATION OF HIRUDINES	ORbRES
A. Mouth with three serrated jaws, or three longitudinal folds.	ARHYNCHOBbELLES
B. Mouth with protractile proboscis.	RHYNCHOBbELLES
CLASSIFICATION OF ARHYNCHOBbELLES	
A. Mouth without jaws but with three longitudinal folds, three or four pairs of eyes, digestive system without gastric caeca, predatory species never hematophagous, no parental protection of cocoons.	ERPOBbELLIbAE family
B. Mouth with three jaws, five pairs of eyes, digestive system with gastric caeca, hematophagous, no parental protection of cocoons.	HIRUbINIbAE family
CLASSIFICATION bES RHYNCHOBbELLES	
A. Mouth equipped with a proboscis, eyes concentrated on the mediodorsal line, dorsoventrally flattened foliaceous body whose organs are visible by transparency, hematophagous or predatory species, presence of parental protection of cocoons and juveniles.	Family SLOSSIPHONIIDAE

The number of eyes, their position and their arrangement are important in determining gender: (Present work)

1 - Four pairs of *eyes....... Theromyzon*

One has three pairs of eyes.............2

2 Three pairs of eyes.... *Alboglossiphonia*

Two pairs of eyes...................3

3 - A single pair of eyes separates one from the other *Batracobdella*

- A single pair of eyes not separated from each other *Placobdella*

IV.GEOGRAPHICAL DISTRIBUTION OF SPECIES ENCOUNTERED

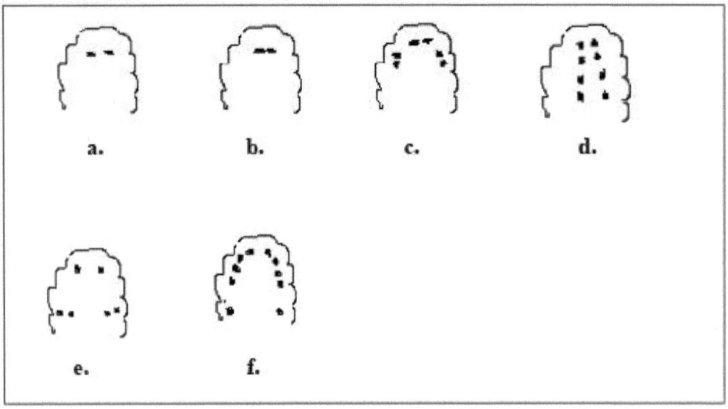

Fig. 76: Position and arrangement of the eyes in the six genera of Hirudinees collected in the Lebna dam: four different genera of the Glossiphoniidae family: (a) *Batracobdella*, (b) *Placobdella*, (c) *Alboglossiphonia*, (d) *Theromyzon*, of the Erpobdellidae family: (e) *Erpobdella*, of the family Hirudinidae: (f) *Hirudo* (Present work).

Following this work, carried out for the first time in Tunisia, we can establish a new Mediterranean and worldwide geographical distribution for some of the species collected.

1. Geographical distribution of *Batracobdella algira*

This species was first discovered by Moquin-Tandon (1846) in Algeria **(Minelli, 1979).** It has also been recorded on the northern shores of the Mediterranean: the Iberian Peninsula, the Balearic Islands, Corsica, Sardinia and also in eastern Europe: Crimea **(Minelli, 1979).** In Tunisia, it was reported by **Blanchard (1908).**

2. Geographical distribution of *Theromyzon tessulatum*

This species was first mentioned by Muller (1774) in Denmark **(Minelli, 1979).** It is known as Holarctic (Palearctic region and Nearctic region). It has also been recorded in Japan and South America **(Minelli, 1979)**.

In the present work it is reported for the first time from Tunisia and North Africa. *T. tessulatum* is the most widely distributed of the other species (Figs. 77 and 78).

3. Geographical distribution of *Placobdella nabeulensis*

P.costata was first discovered in Crimea by Muller (1846) **(Minnelli, 1979)**. It is a

Mediterranean species which extends westwards to the Ukraine. It has also been recorded in Greece, Turkey, the southern part of the Arabian Peninsula and in Italy: Rome, Sassari and Padova **(Minelli, 1979).** In Tunisia, it was recorded as *Haementeria costata* by **Blanchard (1908).**

In this work, we have the opportunity to describe a new species of the genus *Placobdella*, presenting morphological and anatomical similarities with *P. costata*. This is a cryptic species, which can only be distinguished with certainty by molecular analysis.

4. Geographical distribution of *Alboglossiphonia hyalina*

This species was first discovered in Europe by Linnaeus (1761) (**Anonymous 3**: website). It is only known from Central and Eastern Europe (**Anonymous 3**: website). In addition, it has recently been reported from the Indian coast **(Nessemenn, 2002:** website**).**

This is the first time it has been reported in Tunisia and North Africa.

The Glossiphoniidae family is considered cosmopolitan **(Govedich and Davies, 1998)** since it is only found in the Antarctic. In our case, four representatives of this family were found in the Lebna dam.

5. Geographical distribution of *Hirudo troctina*

This species was first discovered by Johonsson (1816) **(Sawyer, 1986).** There is no typical locality. It has been reported by **Hetchel and Sawyer (2002)** in North Africa: Tunisia, Morocco and Algeria and by Utevsky in Ukraine (personal communication).

6. Geographical distribution of *Erpobdella testacea*

This species was first discovered by Savigny (1820) in Paris **(Minnelli, 1979)**. It has also been recorded on the northern shores of the Mediterranean: Spain, northern Italy, Greece and France; eastern Europe; the southern part of Great Britain; Sweden and Finland **(Minnelli, 1979)**.

This is the first time it has been reported in Tunisia and North Africa.

CHAPTER IV

ECOLOGICAL AND BIOLOGICAL STUDY

Few studies have been carried out on the ecological preferences of leeches. In fact, these animals inhabit most aquatic environments and the limiting factors seem to be temperature and high acidity **(Grasse, 1959).** However, these leeches are exposed to extreme variations in environmental factors such as temperature and the oxygen content of the water.

Many leeches are able to tolerate a wide variation in salinity. They have also been found to be highly tolerant of pollution, pesticides, oils and sulphates, while never tolerating zinc (**Ginsberg, 1998: website**). This relative tolerance is explained by the presence of physiological plasticity **(Govedich, 2001).**

During the study period (September 2005-August 2006), we measured the salinity, pH and temperature of the dam water on a monthly basis (Table 3) at the same time as the worm harvest.

The study of these abiotic factors was carried out with the aim of demonstrating a possible relationship between environmental fluctuations and the variation in the frequency of each species in the monthly samples.

Table 3: Abiotic parameters recorded monthly at the study station (Present work).

Month	J	F	M	Av	M	Ju	Jui	Visit	s	O	N	b
Temperature	10,4	9,7	12,8	14,2	16,3	21,1	24,7	21,6	19,6	20,7	18,9	16,1
Salinite	0,71	0,61	0,60	0,59	0,58	0,60	0,75	1,19	1	1,09	0,93	0,81
pH	8,4	7,7	7,4	7,9	8,2	8,6	8,7	8,5	8,1	7,9	8,3	8,2

1. ABIOTIC FACTORS AND THEIR INFLUENCE ON HIRUDINE ABUNDANCE

1. Abiotic factors

1.1. Water temperature

Temperature is an important distribution factor for aquatic life. It controls all the biological processes (reproduction, growth, thermal preference, etc.) linked to a given environment **(Aminot et chaussepied, 1983).**

During the study period, the water in the Lebna dam reservoir is relatively warm in the summer (June, July and August), with peaks in July (24.7°C) and August (21.6°C). The water cools progressively in autumn (September, October and November) and winter (December and January), with minimum values recorded in February (9.7°C) and January (10.4°C) (Fig. 79). These results are in agreement with those found by **Ammar (2006).**

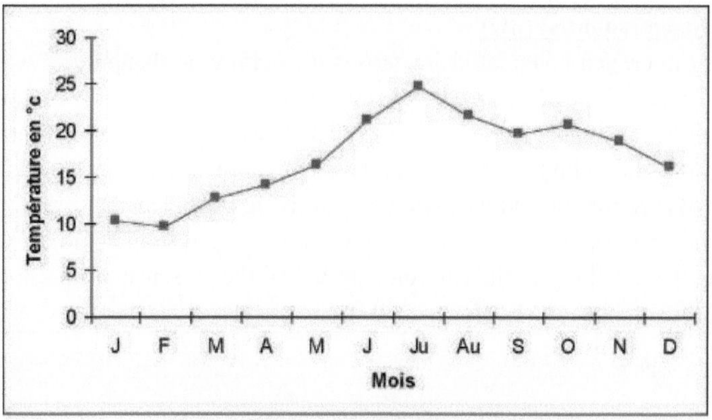

Fig. 79 : Variation mensuelle de la température.

1.2. Salinite

The mineralisation of water depends essentially on the terrain through which it flows **(Nisbet and Verneaux, 1970).**

The waters of the dam are marked by relatively close salinities (Fig. 80). They show temporal fluctuations.

Surface water salinity values recorded during the study period fluctuated between 0.58 g/l and 1.93 g/l. Seasonal variations in salinity were small. They may be due to the effect of evaporation in summer and the inflow of fresh water from precipitation during the cold season.

Compared with the observations of **Dhaouadi (2003)**, at the Lebna reservoir during the period from July 97 to December 98, the maximum salinity reached a value of 0.97g/l. This value is lower than that recorded in the present study. This value is lower than that recorded in the present study.

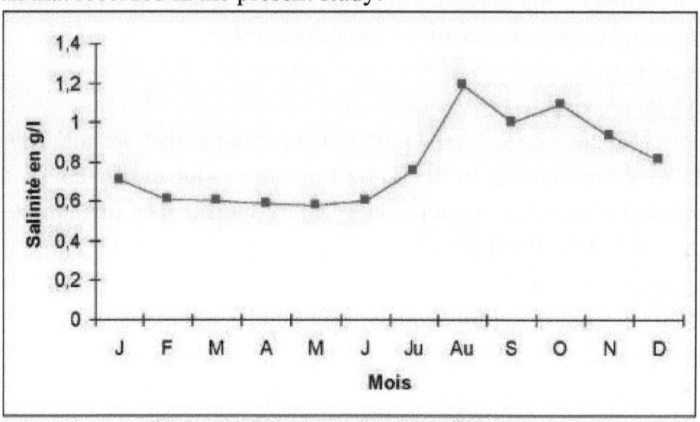

Fig. 80 : Variation mensuelle de la salinité.

3. Hydrogen potential (pH)

The pH or hydrogen potential characterises the acidity or alkalinity of water. This parameter makes it possible to estimate the degree of aggressiveness of a water and its suitability for animal life. According to **Dussart (1992)**, in freshwater, pH is quite variable depending on the time of day, temperature, insolation, light intensity, chlorophyll assimilation and the respiration of living organisms.

The pH values measured during the sampling trips ranged from 7.4 to 8.7 (Fig. 81), indicating the alkalinity of the environment due to the presence of limestone rocks, and therefore carbonates **(El Herry, 2004).**

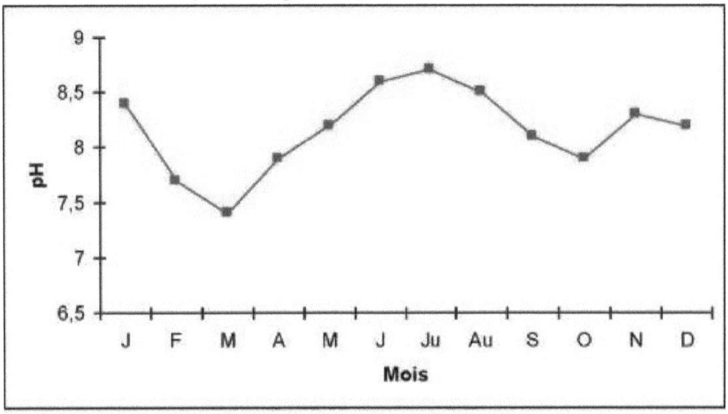

Fig. 81 : Variation mensuelle du pH.

2. Influence of Abiotic Factors on the Numbers of Different Hirudine Species

With the exception of *E. testacea*, monthly measurements of abiotic factors between September 2005 and August 2006 showed a lack of correlation between these factors and the abundance of the species recorded.

2.1. Temperature

Batracobdella algira

According to Figure 82, *B. algira* prefers temperatures that are moderately high, around 15°C, corresponding to the highest numbers. However, analysis using the Spearmann test showed no correlation between the number of this species and temperature (R = 0.49; P = 0.099).

Theromyzon tessulatum

Examination of Figure 83, illustrating the monthly variations in the number of *Theromyzon tessulatum* collected as a function of temperature, shows that this species is absent throughout the spring and summer period, when the highest temperature values were recorded. However, analysis using the Spearmann test showed that there was no correlation between temperature and the abundance *of T. tessulatum* (R = -0.241; P = 0.449).

Placobdella nabeulensis

Figure 84, showing the monthly variations in the number of *P. nabeulensis* collected as a function of temperature, shows that the latter seems to have no influence on the numbers of this species, since for two similar temperatures (June 21.1 and August 21.6), the number of specimens collected in the same area is different (5 and 18). However, it should be noted that at low temperatures below 16° (December, January, February, April and May), the species is not very abundant (N less than 5). Statistical analysis of the results using the Spearmann test confirms this hypothesis. In fact, no correlation was found between temperature and the abundance of this species (R = 0.340; P = 0.278).

Alboglossiphonia hyalina

Figure 85 shows that water temperature appears to have no influence on the presence of *A. hyalina* species. In fact, for almost the same temperature value, there are two different numbers (May: T = 16.3 and N = 35; December: T = 16.1 and N = 4). Applying the Spearmann test, no correlation between temperature and the abundance of this species was found (R = 0.19; P = 0.54).

Hirudo troctina

Observation of the graph in figure 86 shows at first sight that temperature seems to have an effect on the abundance of *H. troctina* since the highest water temperature coincides with the maximum number of specimens collected. However, statistical analysis of the results showed no correlation between temperature and the abundance of this species (R = 0, 280; P = 0, 377).

Erpobdella testacea

Analysis of Figure 87 shows that *E. testacea* is present throughout most of the year, suggesting that the tolerances for this species are wide. The highest density is recorded during the warm season, suggesting that this species prefers high temperature values.

Statistical processing of these data showed a significant correlation at the 5% threshold between temperature and the abundance of this species (R = 0.619; P = 0.031).

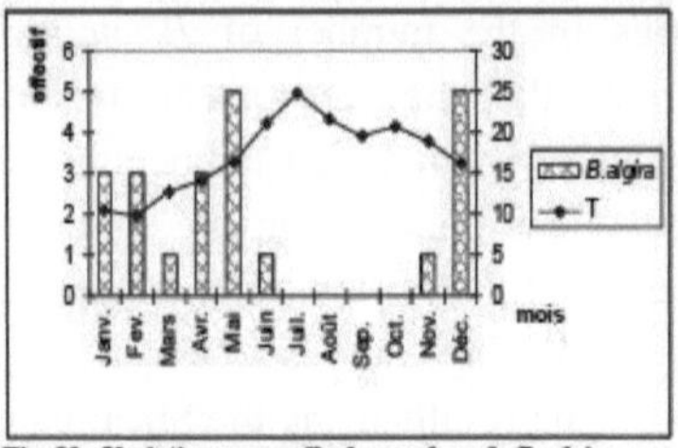

Fig. 82 : Variation mensuelle du nombre de *B. algira* récolté en fonction de la température.

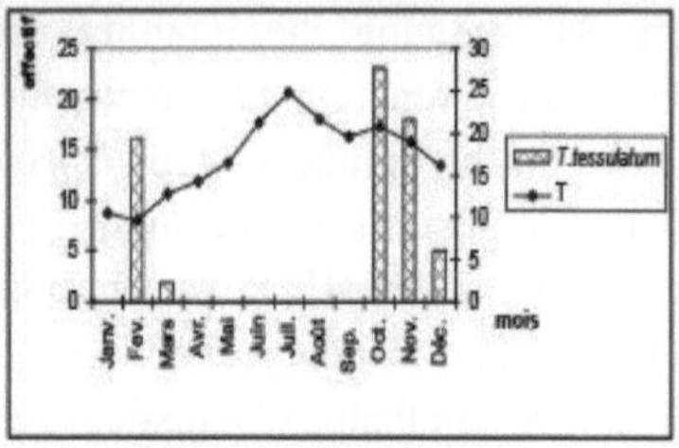

Fig. 83 : Variation mensuelle du nombre de *T. tessulatum* récolté en fonction de la température.

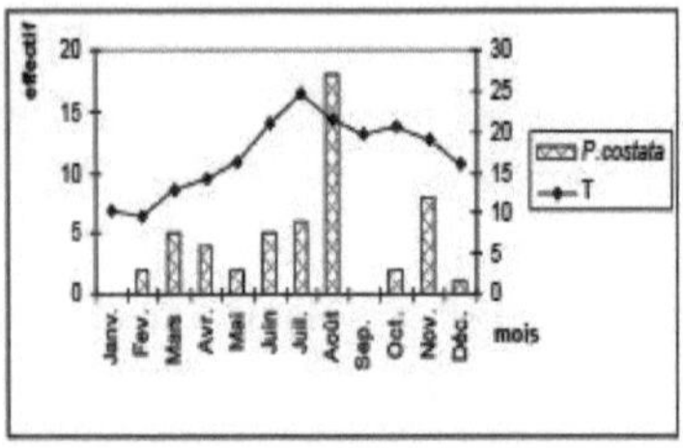

Fig. 84 : Variation mensuelle du nombre de *P. babeulensis* récolté en fonction de la température.

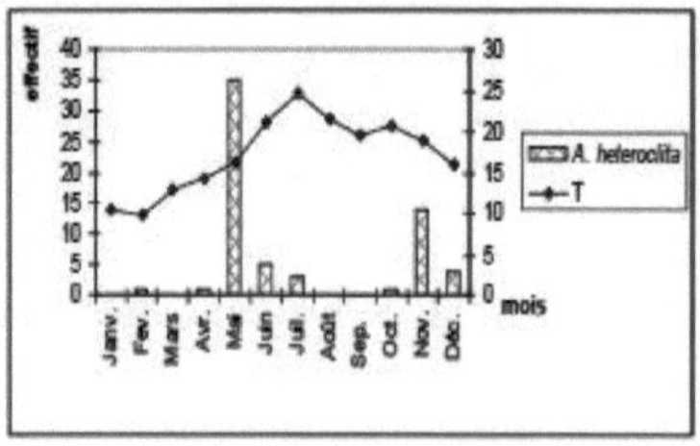

Fig. 85 : Variation mensuelle du nombre d'*A. hyalina* récolté en fonction de la température.

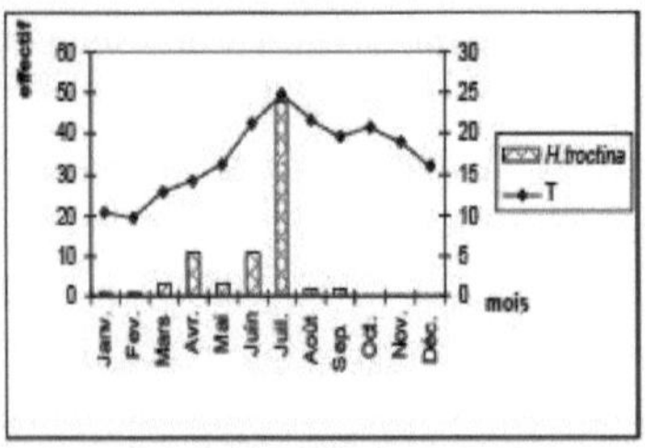

Fig. 86 : Variation mensuelle du nombre de *H. troctina* récolté en fonction de la température.

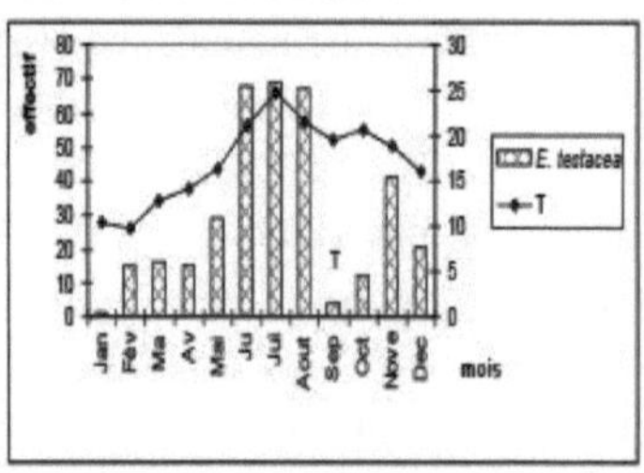

Fig.87 : Variation mensuelle du nombre d'*E. testacea* récolté en fonction de la température.

Fig. 82: Monthly variation in the number of *B. algira* harvested as a function of temperature.
Fig. 83: Monthly variation in the number of *T. tessulatum* harvested as a function of temperature.
Fig. 84: Monthly variation in the number of *P. babeulensis* harvested as a function of temperature.

Fig. 85: Monthly variation in the number of *A. hyalina* harvested as a function of temperature.
Fig. 86: Monthly variation in the number of *H. troctina* collected as a function of temperature.
Fig.87 : Monthly variation in the number of *E. testacea* harvested as a function of temperature.

.2. Salinite

Batracobdella algira

Figure 88, showing monthly monitoring of *B. algira* numbers as a function of salinity, shows that salinity appears to have no influence on the presence of this species. In fact, we recorded variable numbers for almost the same salinity values (February to June). The Spearmann test showed that there was no correlation between salinity and the presence of *B. algira* at the study station (R = 0.088; P = 0.784).

Theromyzon tessulatum

Figure 89, representing the monthly variations in the number of *Theromyzon tessulatum* as a function of salinity, shows that salinity seems to have no effect on this species since the number of fish is the same (November: salinity = 0.93 and N = 18; February: salinity = 0.61 and N = 16). Statistical analysis of the results showed no correlation between the number of *T. tessulatum* and variation in salinity (R = 0.31; P = 0.322).

Placobdella nabeulensis

Monthly variations in the number of *P. nabeulensis* harvested as a function of salinity (Fig. 90) show that salinity seems to have an influence on the numbers of this species. Indeed, it was noted that the maximum salinity recorded coincided with the highest population size. However, statistical analysis of the results invalidated this hypothesis and no correlation between salinity and *P. costata* numbers was found (R = 0.088; P = 0.784).

Alboglossiphonia hyalina

Figure 91 shows that the abundance of *Alboglossiphonia hyalina* appears to be independent of variations in salinity. Different harvests were obtained for the same salinity value (April: salinity = 0.59 and N = 1; May: salinity = 0.58 and N = 35). Statistical analysis of the results confirms this hypothesis and no correlation was found between the abundance of *A. hyalina* and salinity (R = - 0, 332; P = 0, 291).

Hirudo troctina

According to Figure 92 and on the basis of statistical analysis, the abundance of *H. troctina* appears to be independent of the salinity of the water (R = - 0, 557; P = 0, 59).

Erpobdella testacea

Examination of Figure 93 shows that salinity seems to have no effect on the abundance of this species, since the same number of specimens can be found collected for different salinity values (June: salinity = 0.6 and N = 78; August: salinity = 1.19 and N = 79). Analysis using the Spearmann test showed that there was no correlation between the abundance of *E. testacea* and salinity (R = -0.056; P = 0.862).

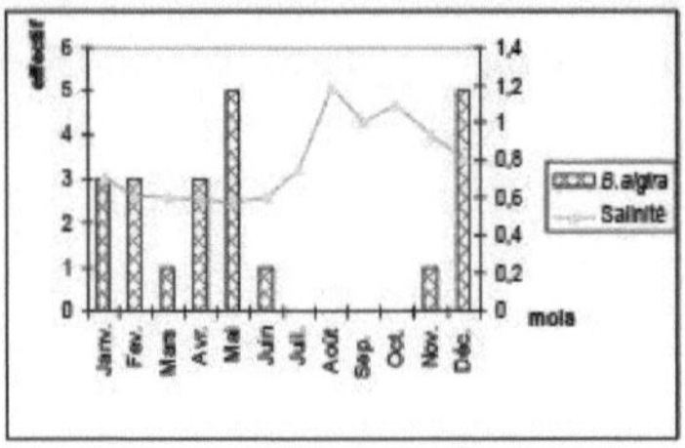

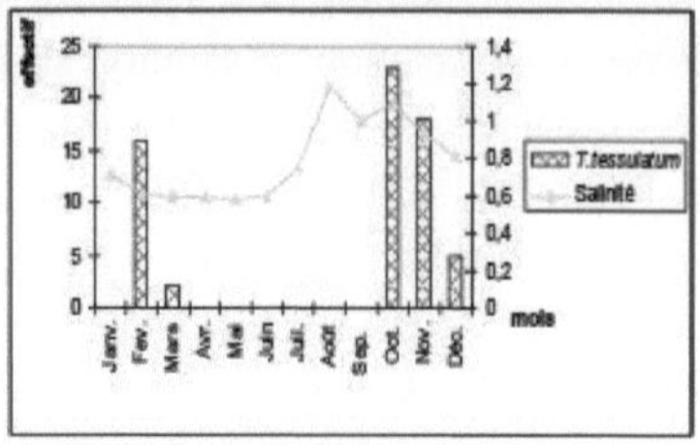

Fig. 88 : Variation mensuelle du nombre de *B. algira* récolté en fonction de la salinité.

Fig. 89 : Variation mensuelle du nombre de *T. tessulatum* récolté en fonction de la salinité.

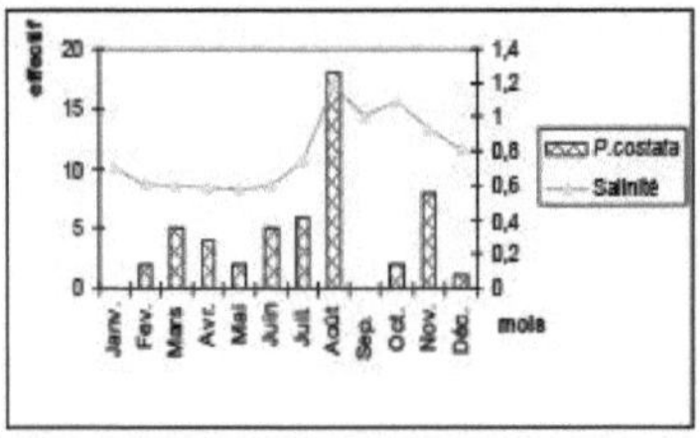

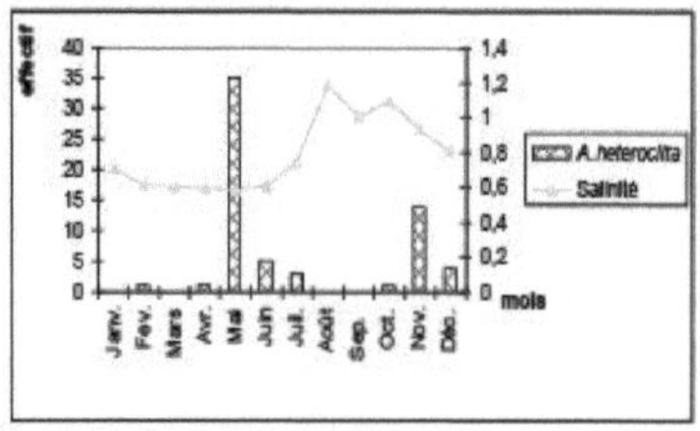

Fig. 90 : Variation mensuelle du nombre de *P. nabeulensis* récolté en fonction de la salinité.

Fig. 91 : Variation mensuelle du nombre d'*A. hyalina* récolté en fonction de la salinité.

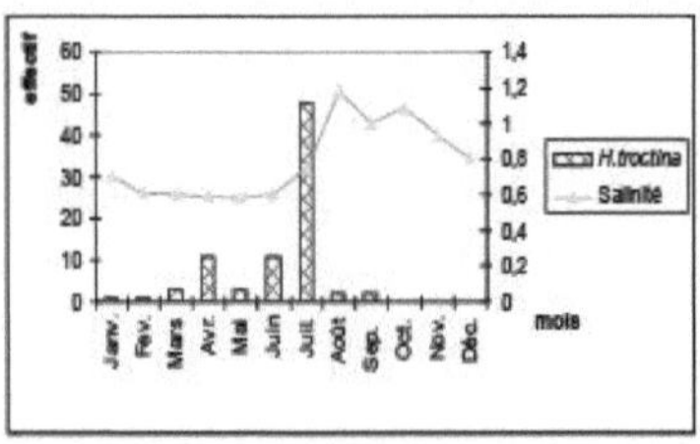

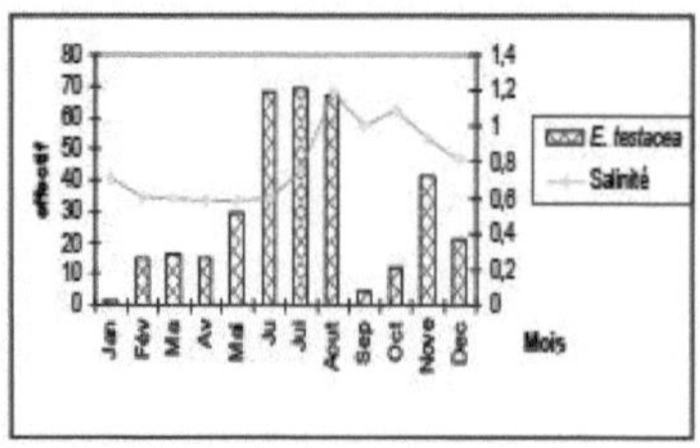

Fig.92 : Variation mensuelle du nombre de *H. troctina* récolté en fonction de la salinité.

Fig. 93 : Variation mensuelle du nombre d'*E. testacea* récolté en fonction de la salinité.

Fig. 88: Monthly variation in the number of *B. algira* harvested as a function of salinity.
Fig. 89 Monthly variation in the number of *T. tessulatum* harvested as a function of salinity.
Fig. 90 Monthly variation in the number of *P. nabeulensis* harvested as a function of salinity.
Fig. 91 Monthly variation in the number of *A. hyalina* harvested as a function of salinity.
Fig.92: Monthly variation in the number of *H. troctina* harvested as a function of salinity.
Fig. 93: Monthly variation in the number of *E. testacea* harvested as a function of salinity.

2.3. The pH

Batracobdella algira

Figure 94, showing the monthly monitoring of the number of *B. algira* collected as a function of pH, suggests that the abundance of this species depends on salinity around a value of ± 8. In addition, we recorded different results for the same pH in October and April. Statistical analysis of the results showed no correlation between *B. algira* numbers and pH (R = 0.340; P = 0.278).

Theromyzon tessulatum

Examination of Figure 95 shows that the absence of *T. tessulatum* coincides with the highest pH values. However, two harvests were recorded that are relatively distinct for two pH values that are practically the same (May: pH = 8.2 and N = 0;

November: pH = 8.3 and N = 23). Statistical analysis of the results showed no correlation between the abundance of *T. tessulatum* and pH (R = -0.489, P = 0.106).

<u>Placobdelle nabeulensis</u>

Figure 96, showing monthly variations in the number of *P. costata* harvested as a function of pH, shows that pH has no influence on this species. In fact, for two identical pH values (November: 8.3 and December: 8.2), the number of individuals harvested is different (N = 8 and N = 1). Statistical analysis of the results confirmed this hypothesis and no correlation was found (R = 0.340; P = 0.784).

<u>Alboglossiphonia hyalina</u>

The pH appears to have no effect on the number of *Alboglossiphonia heteroclita* collected (Fig. 97). This conclusion is drawn from the fact that for pH values of 8.1 and 8.2 found in September and May respectively, two different numbers were recorded (0 and 35). Statistical analysis of the results (correlation test of Spearmann) confirms this hypothesis and no correlation was found between the number of *A. heteroclita* and pH (R = 0.19; P = 0.546).

<u>Hirudo troctina</u>

Although the pH peak coincides with the best harvest, it appears to have no effect on the presence of *H. troctina* (Fig. 98). Analysis by Spearmann's test showed no correlation between *H. troctina* numbers and pH variation (R = 0.280; P = 0.377).

<u>Erpobdella testacea</u>

According to figure 99, the highest harvest was recorded during the months of June, July and August, the months in which the highest pH values were recorded. This proves that the *E. testacea* species prefers water with an alkaline pH.

A statistical analysis was carried out and a statistically significant positive correlation at the 5% threshold was found between the number of *E. testacea* and pH (R = 0.616; P = 0.032).

<u>Conclusion</u>

With the exception of *E. testacea*, whose numbers were statistically correlated with temperature and pH, there was no correlation between the three abiotic factors studied and the numbers of the other Hirudineae species reported.

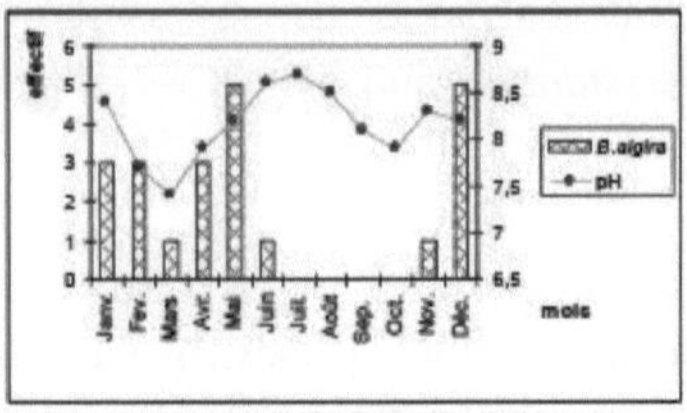

Fig. 94 : Variation mensuelle du nombre de *B. algira* récolté en fonction du pH.

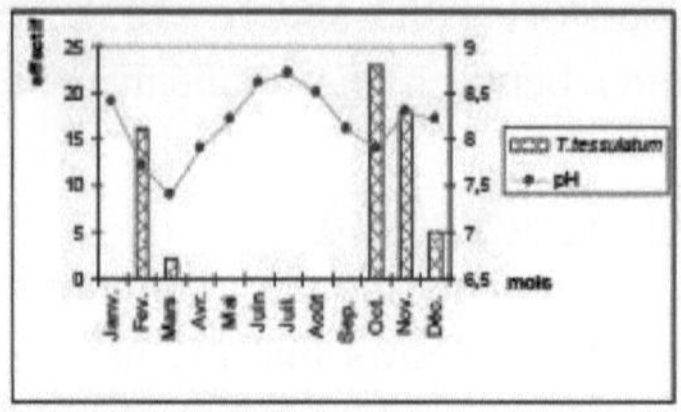

Fig. 95 : Variation mensuelle du nombre de *T. tessulatum* récolté en fonction du pH.

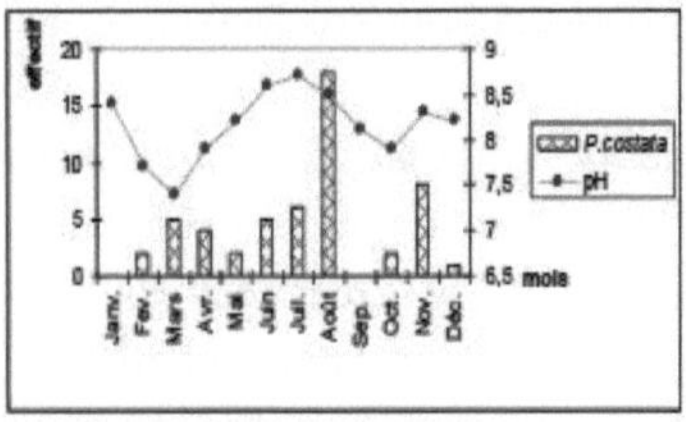

Fig. 96 : Variation mensuelle du nombre de *P. costata* récolté en fonction du pH.

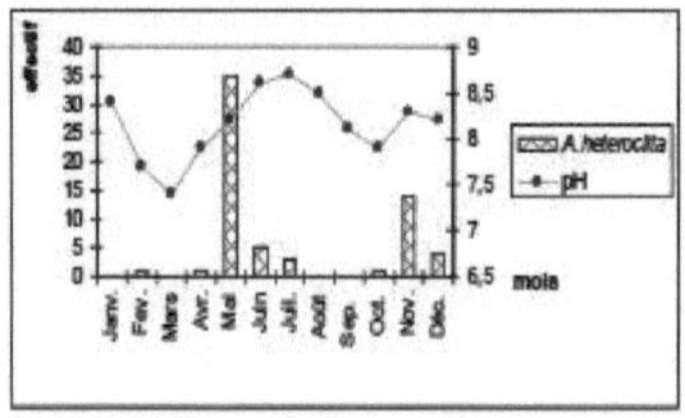

Fig. 97 : Variation mensuelle du nombre d'*A. heteroclita* récolté en fonction du pH.

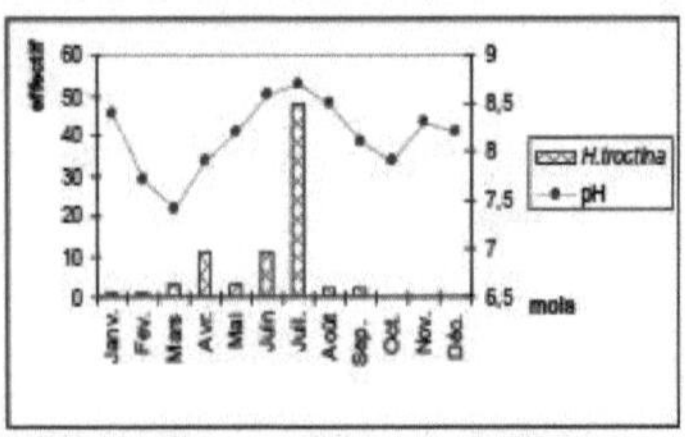

Fig. 98 : Variation mensuelle du nombre de *H. troctina* récolté en fonction du pH.

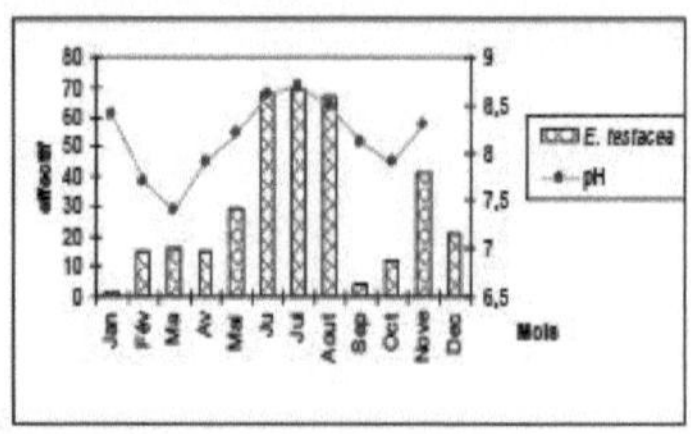

Fig. 99 : Variation mensuelle du nombre d'*E. testacea* récolté en fonction du pH.

Fig. 94: Monthly variation in the number of *B. algira* harvested as a function of pH.
Fig. 95: Monthly variation in the number of *T. tessulatum* harvested as a function of pH.
Fig. 96: Monthly variation in the number of *P. costata* harvested as a function of pH.
Fig. 97: Monthly variation in the number of *A. heteroclita* harvested as a function of pH.
Fig. 98: Monthly variation in the number of *H. troctina* harvested as a function of pH.
Fig. 99: Monthly variation in the number of *E. testacea* harvested as a function of pH.

II. STRUCTURE OF THE POPULATION STUDIED

We collected 683 leeches each month (table 4). It was noted that some taxa fluctuate seasonally and others only occur during certain periods of the year. The structure of the Achetes population varies from one season to another, and even from one month to another.

1. Specific richness

The census revealed six species with certainty (S = 6) and three others that were not identified and could represent other species. The latter are indicated in Table 4 by *Hirudo sp.*

Table 4: Monthly distribution of species found in the Lebna dam

	S. algira	*T. tessulatum*	*P. costata*	*A. heteroclita*	*H. troctina*	*H .sp*	*E. testacea*
September	0	0	0	0	2	0	4

October	0	23	2	1	0	1	12
November	1	18	8	14	0	1	41
December	5	5	1	4	0	1	21
January	3	0	0	0	1	13	1
February	3	16	2	1	1	1	15
March	1	2	5	0	3	0	16
April	3	0	4	1	11	6	15
May	5	0	2	35	3	0	29
Join	1	0	5	5	11	5	68
July	0	0	6	3	48	7	69
August	0	0	18	0	2	4	67
Total	22	64	53	64	82	39	358

2. Relative abundance

Table 5 shows the number of taxa and families in the stand studied, together with their relative abundance.

Firstly, *Erpobdella testacea* is the most common species in the material collected. It represents more than half with a percentage of **52.49%**. The conditions at the dam seem to be favourable for the establishment of this species. However, we have already mentioned that, according to the literature, Erpobdelles are indicators of pollution. According to the present study, the presence of *Erpobdella testacea* is not closely linked to pollution, since the dam water is mainly used for irrigation. The abundance of this species can also be explained by the nature of its diet, since it is a predatory species rather than a hematophagous ectoparasite.

The low numbers of specimens caught from the Glossiphoniidae family **(29.75%)** and the Hirudinidae family **(17.74%)** can be attributed to their diet. They are all temporary hematophagous ectoparasites.

Table 5: Numbers and relative abundance of the various Hirudineae during the study year (2005 - 2006).

Taxa	Number of individuals	Relative abundance
Glossiphoniidae	203	29,75 %
Batracobdella algira	21	3, 22 %
Placobdella costata	53	7,77 %
Theromyzon tessulatum	65	9, 38 %
Alboglossiphonia heteroclita	64	9, 38 %
Hirudinidae	121	17,74 %
Hirudo troctina	82	12,02 %
Hirudo .sp.	39	5,71 %
Erpobdellidae	358	52,49 %
Erpobdella testacea	358	52,49 %
Total	Z 682	Z 100 %

Batracobdella algira, taken as an example from the Glossiphoniidae family, has the lowest percentage (**3.22%**). This can be explained by the fact that this species is attached to its host (amphibians) and is therefore not very abundant on the bedrock. This observation was confirmed in the laboratory, since during a breeding programme for *Bufo mauritanicus,* we were able to collect around ten specimens of *B. algira*, which parasitises this amphibian.

Within the Hirudinidae family, *Hirudo troctina* is the best represented with a percentage of **12.02%**.

3. Seasonal variations in the abundance of species collected

In our study, we noted that certain taxa diminished in numbers or were absent altogether during certain periods (*Theromyzon tessulatum*). This is probably due to reproduction, since in some monotelic species, oviposition is accompanied by histolysis, which leads to the death of the animal. Table 6 below summarises the seasonal variation in the taxa collected.

Table 6: Seasonal variations in relative abundance and mean density of taxa sampled in the Lebna dam.

					Spring		*Summer*				*Autumn*	
	N	Ar	0	N	Ar	0	N	Ar	0	N	Ar	0
S. algira	11	11,82	1	8	7,20	0,72	1	0,28	0,09	1	0,77	0,28
P. costata	3	3,22	0,27	11	9,90	1	29	8,30	2,63	10	7,75	0,90
T. tessulatum	21	22,58	1,90	2	1,80	0,18	0	0	0	42	32,55	3,81
A.heteroclita	5	5,37	0,45	36	32,43	3,27	8	2,29	0,72	15	11,62	1,36
H. troctina	2	2,15	0,18	17	15,31	1,54	61	17,47	5,54	2	1,55	0,18
H .sp.	15	16,12	1,36	6	5,40	0,54	16	4,58	1,45	2	1,55	0,18
E. testacea	36	10,05	3,27	60	16,75	5,45	204	56,98	15,92	57	44,10	5,18
Total	93			140			319			129		

N: total number of individuals harvested.

Ar : Relative abandonment in %.

D: average density (ind/m).[2]

If we examine the seasonal variations in the population of the Lebna dam reservoir (Tab. 6 and Fig. 101), we can see that :

- *Erpobdella testacea* is the most abundant species, with a demographic explosion noted mainly during the summer period (Ar: **56.98%** and d = **15.92**). Its abundance can be explained, as already mentioned, by the fact that it is a predatory species. *E. testacea* feeds mainly on insect larvae **(Kutschera, 2003)**. The latter are abundant throughout most of the year **(Harrath, 1999).**
- During the winter period, in addition to *Erpobdella testacea, Theromyzon tessulatum* was present in high numbers (Ar: **22.58%** and d = **1.90**). The specimens of the latter species are all juveniles, which suggests that a period of egg-laying has taken place. This species is also abundant during the autumn period, but this time it is mainly the mature specimens that appear with their eggs laid on the ventral side of the body (Ar: **32.55** and d = **3.81).** The abundance of this species during these two periods is certainly linked to the abundance of its host (ducks), and this hypothesis is confirmed by specimens captured gorged with blood (Fig. 100).

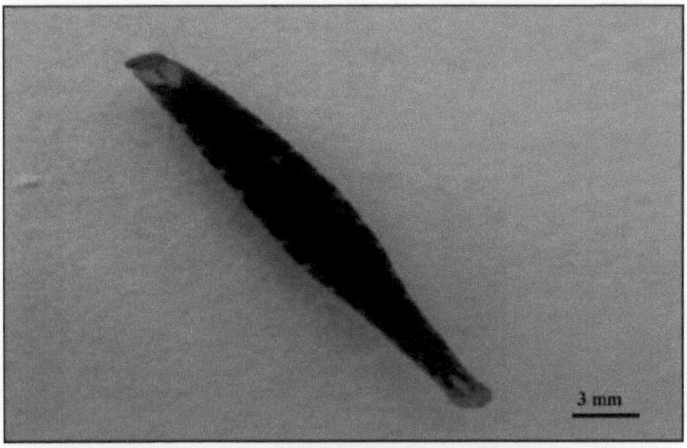

Fig. 100: ***Theromyzon tessulatum*** **blood throat (Present work).**

- *Alboglossiphonia heteroclita* is abundant during the spring period.

(Ar: **32.43%** and d = **3.27**) and autumn (Ar: **11.62%** and d = **1.36)**. The abundance of this species during spring and autumn can be explained by the abundance of Gasteropods at the dam during these two periods **(Harrath, 1999).**

- *Batracobdella algira* is abundant during the winter period (Ar: **11.**82% and d = **1**) as well as during the spring period (Ar: **7.20%** and d = **0.72)**. According to our field observations, these two periods coincide with the abundance of amphibians, the preferred host of this species.
- *Placobdella costata* is present in almost the same abundance during the summer, spring and autumn periods. However, we note that the density is high during the month of August, which is in agreement with the work of **Saglam Naim (2001)** who found that the highest density of *P. costata* was in August (0.45 individuals/m^2) and the lowest (0.05 individuals/m^2) between December and February. This suggests an abundance of prey for this species during the periods mentioned above. This species feeds on the blood of freshwater turtles **(Siddall and Gaffney 2004).** In the present work, it was collected from *Mauremys leprosa.*

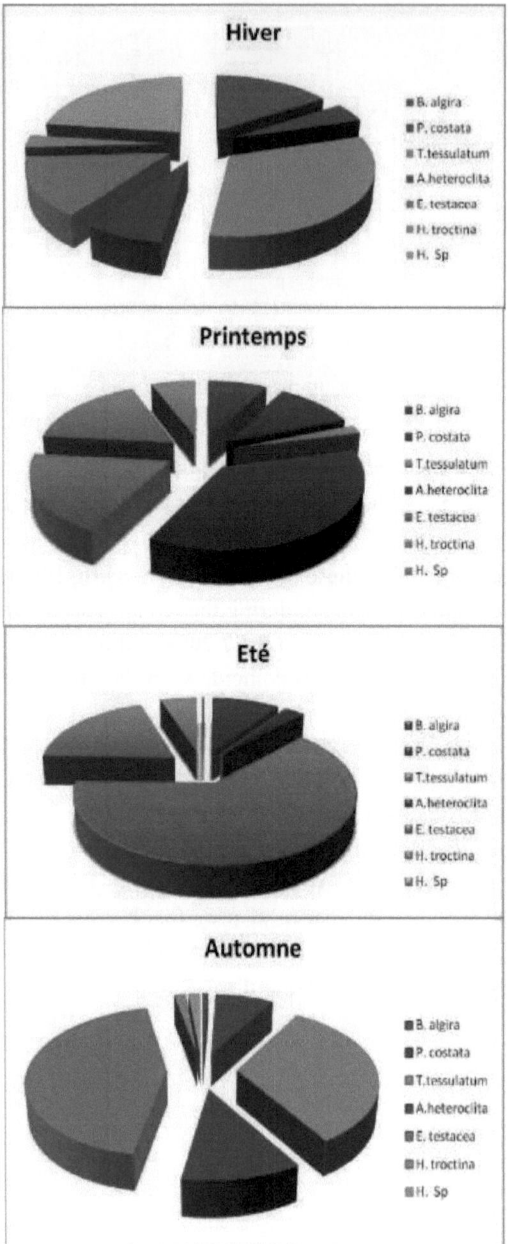

Fig. 101: Seasonal variations in the relative abundance of Hirudinees at the Lebna dam.

III. ADAPTIVE STRATEGIES AND DISCUSSION

Theromyzon tessulatum is abundant during two periods of the year, autumn and winter. The latter period is marked by the exclusive presence of juveniles and no adults are present, whereas during the autumn period, the presence of adults is noted, the majority of which carry their eggs on the ventral side of the body. This observation can be explained by the fact that this species reproduces only once during its lifetime, producing a large number of eggs, after which it dies once the juveniles have been released. This is clearly an "r" stratum with maximum reproductive effort concentrated on a single reproduction (semelparity). This observation has been confirmed in the laboratory and is also consistent with the literature.

Erpobdella testacea is present throughout most of the year with a large number of cocoons. However, these cocoons can be used as food by other species, which probably explains the balance within this population.

The deposition of cocoons on land in *Hirudo troctina* and their tolerance to desiccation can be considered as a completely different strategy to the previous one. This strategy ensures the survival of the species, as the cocoons are thus protected from predators (mainly molluscs), which are mainly found in the water.

IV. BIOLOGICAL STUDY

1. Laying period

Observations in the field and in the laboratory (during rearing), as well as seasonal variations in the average extension length of certain species, have enabled us to suggest the spawning period for the species identified.

In view of the inadequacy of the sampling, only an outline of the life cycles of these species is given in this section. Particular attention is paid to *E. testacea*, for which sampling can be considered sufficient.

The inadequate sampling and lack of harvesting during certain periods of the year could be explained by the migration of species at depth **(Proctor and Young, 1987)** or their attachment to hosts (amphibians, ducks, turtles, molluscs, etc.). Hirudines are in fact temporary ectoparasites.

J Erpobdella testacea

All the monthly samples taken during our study enabled us to collect a total of 358 individuals. Figure 102 shows that of all the specimens collected, 56.98% were collected during the summer and only 43% during the other seasons. This suggests that summer is more favourable for *Erpobdella testacea* activity at the surveyed station.

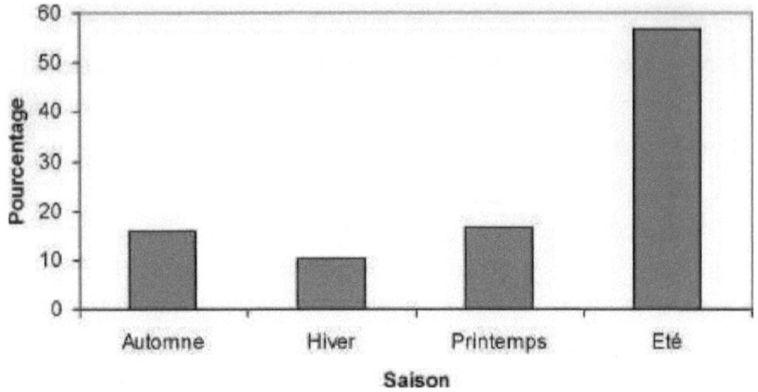

Fig. 102: Percentage of individuals of *E. testacea* harvested according to season.

Figure 103, showing the seasonal variation in mean length of *E. testacea*, shows that during the winter and spring periods, the harvest is made up mainly of specimens of the largest size (4 cm). This suggests that during this period the specimens are sexually mature and able to reproduce.

This finding was confirmed by the fact that throughout this period, the cocoons were fixed under the pebbles. We even monitored the embryonic development of some cocoons, which lasted an average of 40 days at 18°C.

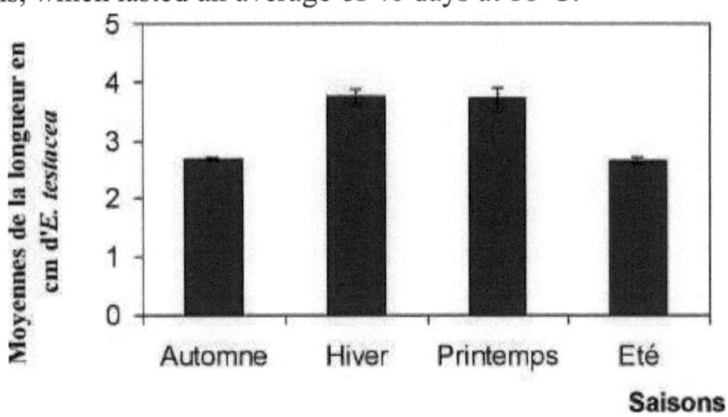

Fig. 103: Seasonal variation in the mean length of *Erpobdella testacea.*

During the summer period, we mainly collected immature individuals ranging in size from 2 to 2.5 cm. The demographic explosion of the population during this period can be attributed to the juveniles from the spring and winter clutches.

J Batracobdella algira

According to our observations, the oviposition period for *B. algira* seems to be in February, March and April. Indeed, during this period we collected a large number of large individuals (2.5cm to 3cm) done presumed to be mature. In addition, these specimens have their eggs stuck to the ventral side of their body.

J Placobdella nabeulensis

The summer period coincides with the egg-laying season for this species, when we collected large specimens up to three cm long.

J Theromyzon tessulatum

Ovigerous specimens are collected in large numbers during the autumn. It is mainly during the months of October and November that we have collected the maximum number of large specimens, up to 3 cm in length.

J Alboglossiphonia hyalina

The largest sizes were recorded during the spring period. It is during this period that specimens bearing eggs on the ventral side of the body have been collected, suggesting that the egg-laying period for this species extends from February to April.

J Hirudo troctina

The egg-laying period for this species probably takes place during the summer. During this season, we collected large specimens that were sexually mature, since histological studies revealed the presence of functional testicles. In addition, during our rearing in the laboratory, we found a cocoon produced by this species during this period.

2. Cannibalism

Cannibalism can be defined as intraspecific predation. It can occur as a result of a number of factors: competitive interactions, mating, hunger, etc. This behaviour has been observed in many species, from eukaryotes to mammals. However, the phenomenon of cannibalism in Hirudineae is still under debate **(Kutschera, 2003)** and no publication referred to this type of study concerning Clitellates (Hirudineae and Oligochetes) before 2005 **(Kutschera and Roth, 2005)**.

Although cannibalism has only been observed *in Nephelopsis obscura* by **Davies and Everett, 1975 *in* Kutschera (2003)** and recently in a population of *Hirudo medicinalis* by **Kutschera and Roth (2005)**, it has also been observed in *Erpobdella testacea*. According to our observations, these Hirudinees placed side by side in the laboratory at a young age show this phenomenon (Fig. 104).

However, **Kutschera (2003)** did not demonstrate this phenomenon in a farmed population of *Erpobdella octoculata*. Moreover, this observation is in agreement with those of other investigators **(Sawyer, 1986)**.

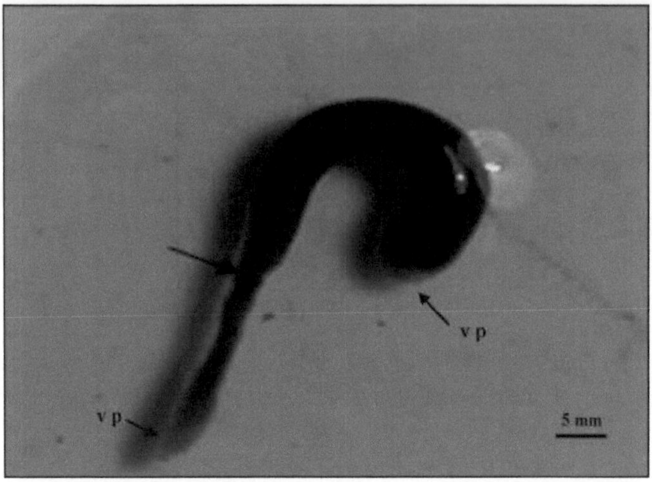

Fig. 104 : Cannibalism in *Erpobdella testacea* (Fleches) v p : posterior sucker (Present work).

In summary, the phenomenon of cannibalism in Hirudinees is still under discussion and has only been noted in two distinct species of Hirudinees: *Nephelopsis obscura* and *Hirudo medicinalis.* In the present work, we report for the first time the phenomenon of cannibalism in *Erpobdella testacea.*

In addition to cannibalism, starving specimens of *Hirudo troctina* have been observed feeding on other specimens of *Erpobdella testacea* (Fig. 105). The latter was in turn observed devouring a specimen of *Theromyzon tessulatum*.

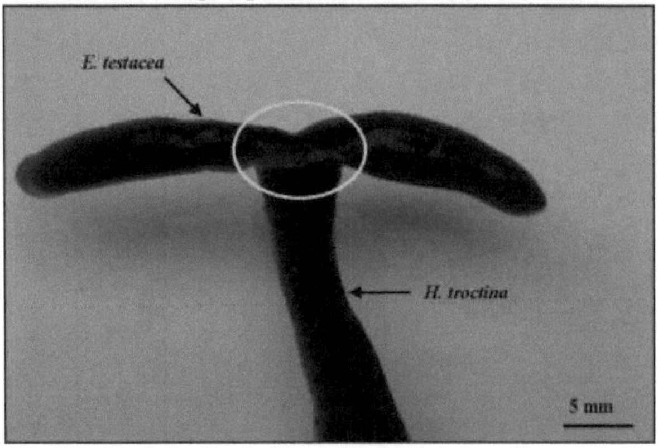

Fig. 105 : *Hirudo troctina* feeding on *Erpobdella testacea* (Present work).

CONCLUSIONS AND OUTLOOK

The study of Annelids Achetes from the waters of the Lebna dam reservoir was carried out in order to identify the species present and to establish the effect of environmental fluctuations on their life cycles.

- The Achetes annelid fauna of the Lebna reservoir is represented by nine species belonging to six different genera and three distinct families: four species of the Glossiphoniidae family have been collected and identified: *Batracobdella algira, Placobdella nabeulensis, Alboglossiphonia hyalina* and *Theromyzon tessulatum*, a single species from the family Erpobdellidae was identified: *Erpobdella testacea* and four species from the family Hirudinidae. Two species of the latter family have been identified (*Hirudo troctina* and *Limnatis nilotica*), while the other three (*Limnatis sp.* 1 and *Limnatis sp.* 2) do not obey any of the keys for determining the different regions described by **Sawyer (1986).** The latter could be species new to science; further molecular study will enable this hypothesis to be confirmed or refuted.
- Among those we have collected, we would like to mention three species that are new to Tunisia and North Africa: *Alboglossiphonia hyalina*, *Theromyzon tessulatum* and *Erpobdella testacea.*
- The phenomenon of cannibalism has been demonstrated for the first time in *Erpobdella testacea.*
- The statistical study revealed a certain correlation between variation in abiotic factors (Temperature and pH) and the abundance of *Erpobdella testacea* in the study station. In fact, it turned out that it prefers slightly warm and alkaline water. All the other species showed no correlation with variations in abiotic factors.
- Based on our observations in the field and in the laboratory, we were able to suggest the life cycles corresponding to some of the species recorded. The oviposition period was limited to the spring period for *Alboglossiphonia hyalina* and *Batracobdella algira*, whereas it was the summer period for *Placobdella nabeulensis* and *Hirudo troctina. Theromyzon tessulatum* seems to have an oviposition period that coincides with the autumn.
- The present work on the Tunisian Hirudinae has shown the presence of a rich and diverse fauna, but we are aware that this inventory is still very sketchy. It represents only a small fraction of the fauna and is only a first step. Further surveys of more diverse biotopes would enable us to enrich it.

In our subsequent studies, we are counting:

> Prospect other regions of Tunisia, or even collaborate with other colleagues in North Africa to establish a determination key for this region.

> Carrying out a DNA-based molecular study to accurately determine the as yet unidentified species of the *Limnatis* genus and establish a phylogenetic relationship between them.

> Study the reproductive biology of the two species *Erpobdella testacea* and *Batracobdella algira* not yet studied.

> To compare the mode of embryonic development of these two species in relation to the cytoplasmic determinants accumulated during ovogenesis.

> A molecular study was carried out to compare the populations of certain Tunisian species, such as *Batracobdella algira* and *Hirudo troctina*, with those reported in Europe.

BIBLIOGRAPHICAL REFERENCES

Aminot A. & Chaussepied M., 1983. Manuel des analyses chimiques en milieu marin. *cnexo, Bndo Documentation Brest*, 395p.

Ammar M., 2006. Contribution a l'etude des copepodes, cladoceres et rotiferes de la retenue de barrage Lebna (Cap bon): Diversité et étude biometrique. *Mastere en Limnologie et ressources vivantes en milieux continentaux.* Institut National Agronomique de Tunis, 133p.

Apakupakul K., Siddall M. E. & Burrreson E. M., 1998. Higher Level Relationships of Leeches (Annelida: Clitellata: Euhirudinea) Based on Morphology and Gene sequences. *Molecular Phylogenetics and Evolution*, **12** : 350-359.

Baker M. W. & Macagno E. R., 2001. Neuronal growth and target recognition: lessons from the leeches. *Can. J. Zoo*, **79:** 204-217.

Beaumont A. & Cassier P., 1981. Biologie animale : des Protozoaires aux Metazoaires epithelioneuriens, I. *Dunod Paris, Bordas Paris*, 459p.

Biernacka B. & Davies R. W., 1994. The effects of temperature and feeding regime on hermaphroditism and gametogenesis of *Nephelopsis obscura*, a freshwater predatory leech. *Can. J. Zool*, **72:** 1639-1642.

Blanchard R., 1891. Short notes on the Hirudinees. I. on the horse leech of North Africa. *Bull. Soc. Zool. Fr.* **16**: 218-221.

Blanchard R., 1917. Monographie des Hemadipsiens (terrestrial leeches). *Bull. Soe. Pathologie Exotique*, **10**: 640-675.

Borda E. & Siddall M. E., 2004a. Arhynchobdellida (Annelida: Oligochaeta: Hirudinida): phylogenetic relationships and evolution. *Molecular Phylogenetics and Evolution*, **30**: 21-225.

Borda E. & Siddall M. E., 2004b. Review of the evolution of life history strategies and phylogeny of the Hirudinida (Annelida: Oligochaeta). *Lauterbornia*, **52 :** 5-25.

Burgin A. M., Eisenhart F. J., Kristan W. B. & French K. A., 2006. Embryonic electrical connections appear to prefigure a behavioural circuit in the leech CNS. *J. Comp. Physiol A.*, **192**: 123-133.

Brumpt E., 1949. Precis de Parasitologie. *Masson Dcle*, 111p.

Denise N. H. & Shankland M., 1992. Lox2, a putative leech segment identity gene, is expressed in the same segmental domain in different stem cell lineages. *Development*, **116** : 697-710.

Dhaouadi S., 2003. Etude ecobiologique d'*Atyaephyra desmarseti* (Millet, 1831) (Crustacea Decapoda, Natania, Atydiae) de trios retenues de barrages de Tunisie: Sidi Salem, Lebna et Sidi Saad. *These de doctorat en sciences biologiques*. Faculte des Sciences de Tunis, 175p.

Doin E. O., 1952. Descriptive Zoology. *Masson Dcle*, 102p.

Dussart B., 1992. Limnologie : l'etude des eaux continentales. *Gauthier-Villars,*

Paris, 681p.

El Herry S., 2004. Caracterisation hydrobiologique et ecotoxicologique des eaux des retenues des barrages du Cap Bon : Bezirk, Chiba, Lebna, Masri et Mlaabi. *Mastere en sciences halieutiques.* Institut National Agronomique de Tunis, 143p.

Esch T. & Kristan W. B., 2002. Decision- Making in the Leech Nervous System. *Integ. And Comp. Biol.* **42**: 716-724.

Figueroa A. O. & Regagnon V. L., 2005. A new freshwater leech species of *Helobdella* (Annelida: Glossiphoniidae) from central Mexico. *Zootaxa*, **976**: 1-8.

Figueroa A. O., Regagnon V. L., & Siddall M. E., 2005. Phylogeny and revision of Erpobdelliformes (Annelida, Arhynchobdellida) from Mexico based on nuclear and mitochondrial gene sequences. *RevistaMexicana de Biodiversidad*, 76 (**2**): 191-198.

Figueroa A. O., 2006. A new species of freshwater leech of the genus *Haementeria* (Annelida: Glossiphoniidae) from Jalisco State, Mexico. *Zootaxa*, **1110**: 39-45.

Govedich F. R. & Davies R. W., 1998. The first record of the genus *Helobdella* (Hirudinoidea: Glossiphoniidae) from Australia, with a description of a new species, *Helobdella papillornata*. *Hydrobiologia*, **389**: 45-49.

Govedich F. R., Blinn D. W., Keim P. & Davies R. W., 1998. Phylogenetic relationships of three genera of Erpobdellidae (Hirudinoidea), with a description of a new genus *(Motobdella*) and species, *Motobdella sedonensis*. *Can. J. Zool.* **76**: 21642171.

Govedich F. R., Blinn D. W., Hevly R. H. & Keim P. S., 1999. Cryptic radiation in Erpobdellid leeches in xeric landscapes: a molecular analysis of population differentiation. *Can. J. Zool*, **77**: 52-57.

Govedich F. R., 2001. A reference guide to the ecology and taxonomy of freshwater and terrestrial leeches (Euhirudinea) of Australasia and Oceania. Cooperative Research Centre for freshwater Ecology Identification Guide No.35.Co-operative Research centre for freshwater Ecology, 67p.

Govedich F. R., Bain B. A., & Davies R. W., 2002. First record of the Asian freshwater leech *Barbronia weberi* (Blanchard, 1897) (Euhirudinea: Erpobdellidae) in Australia. *The Victorian Naturalist*, **119**: 227-228.

Govedich F. R., Burd M., Bain B. A. & Davies R. W., 2003. Reproductive biology of the invasive Asian freshwater leech *Barbronia Weberi* (Blanchard, 1897). *Hydrobiologia*, **510**: 125-129.

Govedich F. R., Negm-Eldin M. M., Bain B. A. & Davies R. W., 2004. Prey preference and feeding behaviour of two Australian Glossiphoniid, and the avoidance response of their snail prey. *The Victorian Naturalist*, **121 (1):** 51-55.

Grasse P. P., 1959. Traite de Zoologie Anatomie Systematique Biologie. *Masson Paris*, **5 (1)**: 471- 593.

Grasse P. P., Poisson R. A. & Tuzet O., 1970. Zoologie 2. Invertebres. *Masson Paris*, 935p.

Grosser C. & Pesic V., 2006. On the Diversity of Iranian Leeches (Annelida: Hirudinea). *Arch. Biol. Sci.* **58 (1)**: 21-24.

Gullo B. S., 1994. Microanatomia de la gonada feminina de *Helobdella triserialis* (Hirudinea, Glossiphoniidae). *anal .soc. Cientif. Argentina*, **224 (2)**: 43-55.

Gullo B. S., 2003. Ciclo reproductor de *Helobdella triserialis* Provincia de Buenos Aires, Argentina. *Rev. Museo La Plata, zoologia.* **15 (167)**: 1-7.

Gullo B. S., 2004. Histologia del sistema reproctor masculino de *Helobdella hyalina* (Hirudinea, Glossiphoniidae) en Argentina. *Iheringia, Ser. Zool, Porto Alegre*, **94 (1)**: 13-18.

Gullo B. S., 2006. *Helobdella nahuelhuapensis* sp. n. (Hirudinea, Glossiphoniidae), from Bariloche, Argentina. *Zootaxa*, **1276**: 33-38.

Harrath A., 1999. Contribution a l'etude des planaires d'eau douce de Tunisie. *DEA d'Ecologie Animate.* Faculte des sciences de Tunis, 151p.

Hetchel F. O. P. & Sawyer R. T., 2002. Toward a taxonomic revision of the medicinal leech *Hirudo medicinalis* Linnaeus, 1758 (Hirudinea: Hirudinidae): redescription of *Hirudo troctina* (Johnson, 1816) from North Africa. *Journal of Natural History*, **36**: 1269-1289.

Kristain W. B., Calabrese R. L. & Friesen W. O., 2005. Neuronal control of leech behavior. *Progress in Neurobiology*, **76:** 279-327.

Kutschera U. & Wirtz P., 1986. A leech that feeds its young. *Animal Behaviour*, **34**: 941-942.

Kutschera U. & Wirtz P., 2001. The Evolution of Parental Care in Freshwater Leeches. *Theory Biosci*, **120**: 115-137.

Kutschera U., 2003. The Feeding Strategies of the Leech *Erpobdella octoculata* (L.): A Laboratory Study. *Internat. Rev.* **88:** 94-101.

Kutschera U., 2004. Species concepts: leeches versus bacteria. *Lauterbornia*, **52**: 171-175.

Kutschera U. & Roth M., 2005. Cannibalism in a Population of the Medicinal Leech (*Hirudo medicinalis*). *Biology Bulletin*, **32**: 626-628.

Kutschera U. & Roth M., 2006a. Notes on the ecology of the Asian medicinal leech *Hirudinaria manillensis* (Hirudinea: Hirudinidae). *Lauterbornia*, **56**: 9-13.

Kutschera U. & Roth M., 2006b. Cocoon deposition and cluster formation in populations of the leech *Hirudo verbana* (Hirudinea: Hirudinidae). *Lauterbornia*, **56**: 5-8.

Mann K. H., 1957. A study of a population of the leech *Glossiphonia complanata. Ibidem*, **26**: 99-111.

Martoja R. & Martoja M., 1967. Initiation aux techniques de 1'Histologie animale. *Masson et cie Paris*, 345p.

Megnin P., 1891. Leeches from Algeria and Tunisia which have remained for more than a month in the mouths of oxen and horses. *Bull. Soc. Zool. Fr.* **16**: 222-227.
Minelli A., 1977. Irudinei: Guide Per Il Riconoscimento Delle Specie Animali Delle Acque Interne Italiane.Consiglio Nazionale Delle Ricerche, 42p.
Minelli A., 1979. Hirudinea. Fauna d'italia. *Calderini Bologna*, 152p.
Moore J. P., 1939. Leeches (Hirudinea) from the Atlas Mountains of Morocco. *Ann.*
Mag. Nat. Hist. **3**: 80-87.
Nesemann H. & Neubert E., 1994. New data to the leeches of the subfamily Trochetinae (Hirudinea, Erpobdellidae). *Miscellanea Zoologica Hungarica*, **9**: 19-28.
Nisbet M. & Verneaux J., 1970. Chemical components of running water. Discussion et proposition des classes en tant que base d'interprétation des analyses chimiques. *Annls Limnol*, **6 (2)**: 161-190.
Oosthuizen J. H. & Davies R. W., 1992. Redescription of the duck leech *Theromyzon rude* (Baired, 1869) (Rhynchobdellida: Glossiphoniidae). *Can. J. Zool*, **70**: 2028-2033.
Paez D., Govedich F. R., Bain B. A., Kellett. M. & Burd M., 2004. Costs of parental care on hunting behaviour of *Helobdella papillornata* (Euhirudinea: Glossiphoniidae). *Hydrobiologia*, **519**: 185-188.
Perret J. L. 1952. Les Hirudinees de la region neuchateloise. *Bull. soc. neuchateloise Sc. Nat.* **75**: 89-138.
Proctor R. M. & Young J. O., 1987. The life history, diet and migration of a lakedwelling population of the leech *Alboglossiphonia heterolicta*. *Hydrobiologia*, **150 (2)**: 133-139.
Sawyer R. T., 1986. Leech biology and behaviour, II. Feeding biology, ecology and systematic. *Oxford University Press New York*, 792p.
Sawyer R. T. & Davies K. D., 1974 - Observations on the physiology and phylogeny of colour change in marine and freshwater leeches (Annelida: Hirudinea). *Hydrobiologia*, **44 (3)**: 215-236.
Saglam N., 2001. First record of the leech *Placobdella costata* (Hirudinoidea: Glossiphoniidae) in Turkey. *Zoology in the Middle East*, **23**: 113-118.
Seaver E. C. & Shankland M., 2000. Leech Segmental Repeats Develop Normally in the Absence of Signals from either Anterior or Posterior Segments. *Developmental Biology*, **224 :** 339-353.
Seaver E. C. & Shankland M., 2001. Establishment of segment polarity in the ectoderm of the leech *Helobdella. Development*, **128**: 1629-1641.
Seurat L., 1922. Fauna of the continental waters of Berberia. Travaux du Laboratoire de Zoologie Appliquee, Faculte des Sciences, Universite d'Alger, 66p.
Shankland M., 1990. Leech Segmentation: Cell Lineage and the Formation of

Complex Body Patterns. *Developmental biology*, **144:** 221-231.

Shankland M., 1994. Leech segmentation: A molecular perspective. *BioEssays*, **16 (11)**: 801-808.

Shankland M., 1995. Formation and Specification of Neurons during the Development of the Leech Central Nervous System. *Journal of Neurobiology*, **27 (3):** 294-309.

Shankland M. & Seaver E. C., 2000. Evolution of the bilaterian body plan: What have we learned from annelids. *BioEssays*, **97 (9)**: 4434-4437.

Siddall M. E., 2001a. Leeches of Laguna Volcan, Bolivia, including a new species of *Helobdella* (Clitellata: Hirudinea). *American Museum Novitates*, **3313**: 1-11.

Siddall M. E., 2001b. Hirudinea from the Apolobampa in the Bolivian Andes, including three new species of *Helobdella* (Clitellata: Hirudinea). *American Museum Novitates*, **3341**: 1-14.

Siddall M. E., Apakupakul K., Burreson E. M., Coates K. A., Erseus C., Gelder S. R., Kallersjo M. & Trapido-Rosenthal H., 2001. Validating Livanow: Molecular Data Agree That Leeches, Branchiobdellidans, and *Acanthobdellida pelidina* from a Monophyletic Group of Oligochaetes. *Molecular Phylogenetics and Evolution*, **21 (3):** 346-351.

Siddall M. E., 2002. Phylogeny of the leech family Erpobdellidae (Hirudinida: Oligochaeta). *Invert. Syst.* **16**: 1-6.

Siddall M. E. & Borda E., 2003. Phylogeny and revision of the leech genus *Helobdella* (Glossiphoniidae) based on mitochondrial gene sequences and morphological data and a special consideration of the *triserialis* complex. *Zoologica Scripta*, **32**: 23-33.

Siddall M. E. & Borda E., 2004. Leech collections from Chile including two new species of *Helobdella* (Annelida:Hirudinida). *American Museum Novitates*, **3457**: 1-18.

Siddall M. E. & Gaffney E. S., 2004. Observations on the Leech *Placobdella ornata* feeding From Bony Tissues of Turtles. *J. Parasitol*, **90 (5)**: 1186-1188.

Siddall M. E., Perkins S. L. & Desser S. S., 2004. Leech mycetome endosymbionts are a new lineage of alphaproteobacteria related to the Rhizobiaceae. *Molecular Phylogenetics and Evolution*, **30**: 178-186.

Siddall M. E. & Budinoff R. B., 2005. DNA-barcoding evidence for widespread introductions of a leech from the South American *Helobdella triserialis* complex. *Conservation Genetics*, **6:** 467-472.

Siddall M. E., Budinoff R. B. & Borda E., 2005. Phylogenetic evaluation of systematic and biogeography of the leech family Glossiphoniidae. *Invertebrate Systematics*, **19**: 105-112.

Soos A., 1966. Identification key to the leech (Hirudinoidea) genera of the world with a catalogue of the species. III. Family: Erpobdellidae. *Acta Zool. Acad. Scient.*

Hung. **12**: 371-407.

Soos A., 1967. Identification key to the leech (Hirudinoidea) genera of the world with a catalogue of the species. IV. Haemadipsidae. *Acta zool. Acad. Scient. Hung.* **13**: 417-43.

Soos A., 1968. Identification key to the species of the genus *Erpobdella* de Blainville, 1818 (Hirudinoidea: Erpobdellidae). *Ann. Hist. nat. Mus. Nat. Hung.* **60**: 141 - 145.

Soos A., 1969. Identification key to the leech (Hirudinoidea) genera of the world with a catalogue of the species.VI: Glossiphoniidae. *Acta zool. Acad. Scient. Hung.* **15**: 397-454.

Zribi S., 2005. Contribution a l'etude de l'ecologie et de la reproduction de *Perinereis Macropus* du Golfe de Gabes (Annelide, Polychete). *Mastere d'Ecologie et Biologie des Populations.* Faculte des sciences de Tunis, 106p.

Trontelj P. & Utevsky S. Y., 2004. Celebrity with a neglected taxonomy: molecular systematics of the medicinal leech (genus *Hirudo*). *Mol Phylogenet Evol*, **34**: 616-624.

Utevsky S. Y. & Trontelj P., 2005. A new species of the medicinal leech (Oligochaeta, Hirudinida, *Hirudo*) from Transcaucasia and an identification key for the genus *Hirudo*. *Parasitol Res*. **98:** 61-66.

Viguier C., 1879. Anatomie comparee des Hirudinees (organisation de la *Batrachobdella (B. Latasti)*). *C. R. Acad. Sci. Paris*, **89:** 110-112.

Weisblat D. A. & Huang F. Z., 2001. An Overview of Glossiphoniid Leech development. *Can. J. zool,* **79**: 218-232.

Westergren S. & Siddall M. E., 2004. Two New Species of Salifid Leeches (Arhynchobdellida: Erpobdelliformes: Salifidae) from South Africa and Madagascar. *American Museum Novitates*, **3456**: 1-6.

ELECTRONIC REFERENCES

Anonymous 1: Faculty.evansville.edu/... ./Annelida/Sld018.htm

Anonymous 2: Nasal leeches. http:// www.nwhc.usgs.gov/publications/ Field-manual.

Anonymous 3: Glossiphoniidae Appendix: http://research.amnh.org/~siddall/glossapp.html **Ginsberg D**., **1998.** Natural History of leeches. http:// www. personal. umich. Edu /~ davegins /leech.html

N.A.S.A., 2006. National Aeronautics and space Administration. World Wide Web electronic publication: http:// zulu.ssc.nasa.gov.

Nessemenn, 2002. http://freenet-homepage.de/hirudinea/Neues.htm

MIX
Papier aus verantwortungsvollen Quellen
Paper from responsible sources
FSC® C105338

Printed by Books on Demand GmbH, Norderstedt / Germany